$20⁰⁰
9/95

RECYCLING and the POLITICS of URBAN WASTE

RECYCLING and the POLITICS of URBAN WASTE

Matthew Gandy

St. Martin's Press
New York

© Matthew Gandy, 1994

First published in the United Kingdom in 1994 by Earthscan
Publications, 120 Pentonville Road, London N1 9JN

First published in the United States of America in 1994

Printed by Clays Ltd, St Ives Plc

ISBN 0-312-12203-9   (cloth)
ISBN 0-312-12204-7   (paper)

Library of Congress Cataloging in Publication Data applied for.

Contents

List of illustrations

FIGURES

TABLES

Glossary

APME	European Plastics Manufacturers Association
BUND	Bund für Umwelt und Naturschutz Deutschland
CCT	Compulsory Competitive Tendering
CDU	Christliche Demokratische Union
DSD	Duale System Deutschland
ERRA	European Recovery and Recycling Association
FDP	Freie Demokratische Partei
FRAPNA	Rhône-Alpes Nature Protection Federation
GAL	Grüne Alternative Liste
HDPE	High-density polyethylene
HEW	Hamburger Electricitätswerke
INCPEN	Industrial Council on Packaging and the Environment
IPCC	Intergovernmental Panel on Climate Change
LAWDC	Local Authority Waste Disposal Company
LB-HSR	Landesbetrieb Hamburger Stradtreinigung
MBI	Market-based policy instruments
NFFO	Non Fossil Fuel Obligation
OECD	Organization for Economic Cooperation and Development
ÖTV	Öffentlicher Dienste, Transport und Verkehr
PET	Polyethylene terephthalate
PVC	Polyvinyl chloride
SELCHP	South East London Combined Heat and Power Ltd
SELWDG	South East London Waste Disposal Group
SPD	Sozialdemokratische Partei Deutschlands
VGK	Verwertungsgesellschaft für gebrauchte Kunstoffverpackungen
VKR	VEBA Kraftwerke Ruhr
WALHI	Indonesian Forum for the Environment

Acknowledgements

Many people have helped me in completing this book, I would like in particular to mention Michael Hebbert, Chris Fernandez, Peter Wicks and Yvonne Rydin at the London School of Economics, Jeff Cooper of the London Waste Regulation Authority, Mike Newport of the Jamestown Road Recycling Centre, Gerd Eich of the Landesbetrieb Hamburger Stadtreinigung and Patricia Grayson in the Department of Sanitation, New York City. This study would not have been possible without the financial assistance I received from the Economic and Social Research Council, the Erasmus Programme and the Nuffield Foundation. I would also like to acknowledge the permission of the Greater London Photograph Library for using their photograph for the front cover and also to Avebury Publishing for permission to reproduce material from my earlier work *Recycling and waste: an exploration of contemporary environmental policy*. Finally, thanks to Susan Rowland at the University of Sussex for assistance with the preparation of maps for Hamburg and New York City and a special thank you to Maria, Billie, Mike, Todd and Crish.

Matthew Gandy
Brighton, England
January 1994

Introduction

This book is about affluence and waste in western economies. It examines the prospects for limiting current patterns of energy and resource use through the recycling of municipal household waste. Three major cities are chosen – London, New York and Hamburg – to illustrate the particular problems posed by attempts to reduce the size of the waste stream in order to protect the environment and achieve more equitable patterns of global consumption.

The twentieth century and particularly the period since World War II has seen a dramatic increase in the production of waste, reflecting unprecedented global levels of economic activity. Figures on primary energy usage in the 1970s indicated that oil consumption in the period 1960–1970 in developed economies was equivalent to total oil produced before 1960 and that coal consumption since 1940 exceeded all the coal produced in the preceding nine centuries.[1] The US alone consumed more minerals from 1940 to 1976 than total global consumption before 1940[2] and the burden of per capita resource use is such that the average US citizen uses some 12,000 tonnes of coal equivalent a year compared with just 55 lbs by the average Ethiopian. The richest 25 per cent of the world population now account for 80 per cent of global energy use and consume 85 per cent of global chemical production and 90 per cent of global automobile production.[3]

On one estimate for the US, municipal wastes have increased five times as quickly as the population over the fifty year period from 1920 to 1970.[4] This increase in the municipal waste stream of western economies can be attributed to a number of factors: rising levels of affluence; cheaper consumer products; the advent of built-in obsolescence; the proliferation of packaging; changing patterns of taste and consumption; and the demand for convenience products.[5] It is not simply the growth of the waste stream and the record levels of consumption for raw materials and energy that has raised concern: there is the environmental impact of the disposal of waste through the use of landfill and incineration; the escalating costs of waste collection and disposal; and the changing composition of municipal waste with greater quantities of toxic materials derived from a variety of products such as paint, solvents and dry-cell batteries.

Since the 1960s the reduction of waste through the promotion of recycling has steadily worked its way up the environmental policy agenda and was endorsed at the 1992 UN Rio Conference where the Agenda 21 declaration called for the promotion of 'sufficient financial and technological capacities at the regional, national and local levels as appropriate to implement waste reuse and recycling policies and actions'.[6] Yet recycling is more than just a response to the environmental crisis and has assumed a symbolic role in beginning to change the nature of western societies and the culture of consumerism. Indeed many environmentalists assume that there will be an inevitable shift from our 'throwaway' society to a post-industrial 'recycling' society of the future, as predicted by the Washington based Worldwatch Institute:

> In the sustainable, efficient economy of 2030 waste reduction and recycling industries will have largely replaced the garbage collection and disposal companies of today...countries will move toward comprehensive systematic recycling of metal, glass, paper and other materials, beginning with source separation at the consumer level.[7]

The prospects for sustainable waste management must be placed in the context of the inexorable trend towards urbanization: in 1950 some 600 million people lived in cities world-wide but by the late 1980s this figure exceeded 2 billion.[8] This is significant because the difficulties in municipal waste management are particularly acute in urban centres as a result of several factors:

- Higher per capita waste generation rates as a result of the greater use of pre-packaged and convenience foods along with higher levels of personal consumption than in rural areas.
- Logistical difficulties in waste collection from high rise apartments where there is a lack of space within the home for the storage of recyclable materials.
- Households living in rooms or apartments often do not have access to gardens for the composting of their putrescible wastes.
- Higher waste collection costs where wastes are collected via communal chutes and paladins from high rise housing.
- Higher waste disposal costs because of dwindling supplies of landfill space and ever greater distances for waste to be transported, necessitating the implementation of complex bulk haulage arrangements; the construction of waste transfer stations and other waste management infrastructure.
- The limited urban space over which recycling facilities such as bottle banks, composting plants and recycling centres must compete with other land uses.

The practical difficulties facing urban waste management are further compounded by the financial difficulties facing many urban municipalities since the 1970s,[9] leading to political conflict between environmentalist demands for comprehensive recycling strategies to tackle the growing waste stream and attempts by city administrations to control their expenditure. I have chosen to focus this book on the cities of London, Hamburg and New York because these cases better illustrate the full extent of the difficulties in promoting sustainable approaches to municipal waste management than do affluent small towns where comprehensive recycling programmes have often been far easier to implement.[10]

I hope that this book will provide more than just a review of current approaches to urban recycling and waste management since the recycling of municipal household waste allows a consideration of three broader themes in contemporary environmental policy making in developed economies:

1. The current emphasis on individual action for environmental protection, by which the ultimate responsibility for protecting the environment is seen to rest with individual citizens rather than Government.[11]
2. The promotion of market-based policy instruments for environmental protection and the view that the environmental crisis is attributable to various forms of 'market failure' because the real costs of environmentally damaging activity are not reflected in the 'price signals' determining current patterns of production and consumption.[12]
3. The promotion of 'sustainable cities' as an integrated approach to urban policy, where the conservation of materials is seen as a fundamental component of environmental protection.[13]

This book is focused on the changing political context for recycling, and in particular the impact of the shift towards more market orientated patterns of public policy since the 1970s. I show that these recent trends in public policy will not lead to higher levels of materials recycling or facilitate the reduction of waste at source in the production process. Contrary to the expectations of many environmentalists, the pattern of municipal waste management is moving inexorably towards the profitable capital intensive option of incineration with energy recovery. This trend is actively promoted by the influential packaging lobby and by the fast-growing private sector waste management industry bidding for lucrative waste collection and disposal contracts. My main conclusion is that any reversal of current trends in order to reduce the size of the waste stream at source would imply not simply a transformation of recycling and waste management policy but also wider shifts within public policy towards a more active role for government in environmental protection at local, national and international levels.

1

The management of municipal waste

Municipal waste accounts for only a relatively small fraction of total global waste production, the main sources being from agriculture, industry and mining. Although this book is focused on wastes produced by individual households it is important to remember that waste and pollution are associated with the manufacture of products and packaging throughout the whole cycle of the primary extraction of materials, the production process, distribution, retail, consumption and in final waste disposal. In Figure 1.1 the contemporary pattern of municipal waste management in developed economies is illustrated, showing how the recycling of post-consumer waste forms part of the wider practice of waste management.

Although municipal waste makes up only a small fraction of the total waste stream many household items such as food, newspapers and electrical goods involve the creation of pollution and waste at earlier stages in the production cycle. In the case of the UK in the late 1980s, some 700 million tonnes of waste were produced (excluding air pollution emissions). Figure 1.2 shows that of this total, 37 per cent comprised agricultural wastes, disposed of as straw and silage; mineral wastes accounted for 34 per cent, disposed of on land; industrial wastes made up 11 per cent of the total and were generally disposed of by landfill or incineration; sewage waste made up 4 per cent of the total, and was mainly discharged to sea; a further 4 per cent was derived from construction and demolition activity disposed of by landfill; 5 per cent consisted of dredged spoil dumped at sea; and finally, municipal wastes amounted to just 5 per cent of the total waste stream, and were disposed of by landfill or incineration.[1] The municipal waste stream includes not only household waste but also comprises quantities of waste derived from other sources such as shops, offices, hospitals and street cleansing. Figure 1.3 shows that the average composition of household waste by weight in developed economies is around 30 per cent paper and cardboard; 24 per cent putrescible kitchen and garden wastes; 11 per cent glass; 9 per cent dust and cinders; 8 per cent plastics; 8 per cent metals; 5 per cent textiles; and a further 4 per cent of miscellaneous items. These categories can be further differentiated: in the case of metals, these are derived mainly from ferrous sources but around 1 per cent of household waste is composed of aluminium. As for plastics there are a variety of different polymers in use, the most frequent types for packaging being HDPE (high-density polyethylene) and PET (polyethylene terephthalate).[2]

Production

Consumption

Waste
management

Environmental
impacts of
waste disposal

Energy recovery
from waste

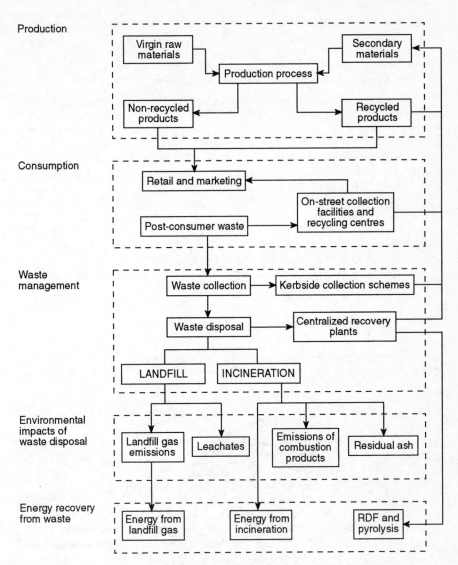

Figure 1.1 Municipal waste management in a developed economy

THE COLLECTION OF MUNICIPAL WASTE

The collection of municipal waste is a typical service provided by local government, which either directly employs labour for the task or uses private sector companies on a contractual basis. Since the 1970s, the use of private companies for the collection of waste has been increasing in developed economies and these changes form an important element in the changing political context for recycling. The technical and logistical aspects of waste collection vary in a number of respects:

- the design of the dustbins and the collection vehicles;
- the time and frequency of the collection service; and
- the particular arrangements used for congested urban areas.

A key dimension to waste collection is the administrative and organizational relationship between the collection and disposal of waste and the interrelationship between the practice of waste management and the structure of local government. In London, for example, there have been 33 separate authorities for waste collection and sixteen for waste disposal since the abolition of the Greater London Council in 1986, whereas in Hamburg and New York both the collection and disposal of waste have been undertaken by one city-wide tier of government throughout the post-war period. A further theme of interest is the rise of political concern, since the mid-1970s, over levels of state expenditure and the cost of local government services, focusing attention on the economic efficiency and productivity of waste collection and forming an important factor behind the privatization of local government services.[3]

WASTE DISPOSAL BY LANDFILL

Throughout human history wastes have been discarded on tips and rubbish dumps but the modern 'landfill site' has emerged in response to the

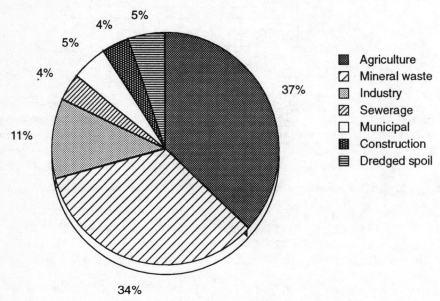

Figure 1.2 Total waste production in the UK

Source: UK Department of the Environment (1992)

recognition of the environmental impact of uncontrolled waste disposal. Previous landfill methods, particularly the mixed disposal of household waste and miscellaneous hazardous wastes derived from industry, were widespread in developed economies until the mid-1970s and this practice has contributed to public opposition towards the construction of new landfill facilities.[4] The use of landfill has been increasingly criticized as a viable disposal option because it produces toxic leachates which can contaminate water supplies and also produces combustible landfill gas from the anaerobic decomposition of putrescible waste. Landfill gases, once they have entered the atmosphere, act as greenhouse agents. One kilogram of methane is believed to be up to 60 times more powerful than carbon dioxide in its contribution to global warming, and methane is thought to constitute some 16 per cent of the greenhouse effect.[5] It has been estimated that landfill gas emissions account for 21 per cent of UK methane gas emissions, being the third most important source after agriculture (33 per cent) and coal mining (29 per cent).[6]

In the US public pressure and new legislation have resulted in a rapid decline in the number of landfill sites: in 1979 there were over 16,000 landfill sites in operation, but by 1988 the number had decreased to around 5500 as a result of tighter controls and the exhaustion of existing sites.[7] The cost of landfill has been rising steadily in many developed economies since the early 1970s[8] and in the US the average cost of landfill per ton rose from $10.80 to $26.93 in the six year period 1982–88 alone.[9] A recent

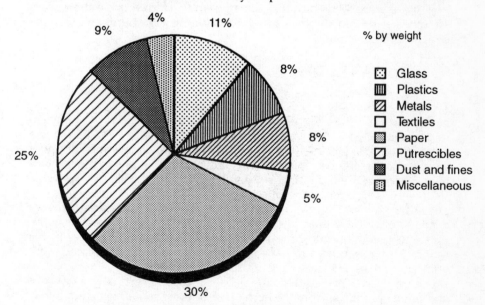

Figure 1.3 The average composition of household waste in developed economies

Source: Cointreau et al (1984), *Recycling from Municipal Refuse: A State-of-the-Art Review*, World Bank, Washington, DC

study for the UK government predicts that landfill costs are set to rise by between 37 and 135 per cent by the year 2000 as a result of the more stringent licensing conditions even without the mooted introduction of a landfill levy.[10] Under the EC Landfill Directive the co-disposal of hazardous and non-hazardous waste would be banned thereby contributing to the rising costs of landfill. A further development is the European Commission's *Green Paper on Civil Liability* which will ultimately make the operators of landfill sites financially liable for any future environmental impacts such as contamination of ground water.[11] In the case of the UK, a survey by Friends of the Earth in early 1993 has revealed that 25 per cent of landfill licences have been surrendered to local government waste regulation agencies as companies seek to avoid post-closure liabilities under new legislation which ensures that site operators pay the full costs of monitoring and controlling landfills until they no longer pose a risk to human health or the environment.[12]

The consequence of these developments is that the new generation of landfill sites differs from the sites of the past: they are larger and further away from centres of population; they are much more closely monitored; their operation often involves advanced technical methods of pollution control; and some sites extract landfill gas as a source of fuel, serving as an additional income for the site operators. Of the 242 landfill gas schemes which are operating world-wide, 55 per cent of projects generate electricity. The US was the first country to exploit landfill gas in the 1970s, and by 1989 there were 87 operational schemes. In western Europe there were 140 schemes running by the late 1980s, including 74 in Germany, 33 in the UK and 20 in Sweden.[13]

WASTE DISPOSAL BY INCINERATION

Technologies to convert waste into steam and electricity were pioneered in Germany and the UK and the first systematic incineration of municipal refuse was tested in Nottingham, England in 1878.[14] By 1912 there were some 76 combined incinerators and electricity works in operation or under construction in the UK and about 17 similar installations in other countries. However the use of incineration began to decline sharply from the 1920s and 1930s onwards because of competition from cheaper waste disposal by landfill. In addition to the economic disadvantages of incineration, the 1960s and 1970s saw criticism of the associated air pollution – in the US alone some 70 per cent of incineration plants in the mid-1970s were judged to have inadequate pollution control and were found to be a significant source of urban air pollution.[15]

Concern over incineration emissions is focused principally on dioxins and furans produced from the burning of chlorine containing compounds such as plastics and bleached paper. Dioxins are among the most toxic molecules yet identified, and research has indicated alarming levels of these carcinogens in mothers' milk in the vicinity of incineration plants.[16]

In 1986, Sweden became the first country to issue dioxin regulations and by 1987 there were 11 air pollutants regulated in West Germany. Further concerns focus on the contribution of these plants to the formation of acid precipitation from the emission of nitrous oxides released with the burning of organic wastes.[17] The other main impact is from the residual ash, containing toxins such as heavy metals left after burning. As a result of the concentration of heavy metals in the incineration ash this is now treated as hazardous waste in Sweden[18] and there are moves in the US to prevent the co-disposal of incineration ash with municipal waste at landfill sites, which could increase the cost of incineration in comparison with alternative means of waste disposal.[19] Table 1.1 shows the development of increasingly tough emissions standards across western Europe in the late 1980s, resulting in the decommissioning and modification of dozens of plants in order to meet higher air quality standards.

Yet the 1980s and 1990s have seen resurgence of interest because of a combination of improved design, enhanced levels of profitability and a political consensus in favour of the expansion of non fossil-fuel sources of energy. The current extent of incineration is illustrated in Table 1.2 and it is now the dominant method of waste disposal for a small number of developed economies, notably in Denmark, Japan, Luxembourg, Sweden and Switzerland. There are now over 500 incinerators for municipal solid waste within the European Community, of which more than 80 per cent have energy recovery facilities.[20] There have been improvements in incineration technology since the 1970s with research and development

Table 1.1 Emissions standards in Europe for waste incinerators

Pollutant mg m^{-3}	France 1986	Germ. 1986	Sweden 1986	EC 1989	Germ. 1989	Neth. 1989
Carbon monoxide	100	100	100	100	50	50
Dust	50	30	20	30	10	5
Sulphur dioxide	–	100	–	300	50	40
Nitrogen dioxide	–	500	–	–	100	70
Hydrogen chloride	100	50	100	50	10	10
Hydrogen fluoride	–	2	–	2	1	1
Chlorine compounds	–	20	–	20	10	10
Inorganic Class 1	0.3	0.2	0.08	0.2	0.1	0.05
Inorganic Class 2	1.0	1.0	–	1.0	0.1	1.0
Inorganic Class 3	5.0	5.0	–	5.0	0.1	1.0

Note: Inorganic Class 1 includes Cadmium and Mercury
Inorganic Class 2 includes Arsenic; Cobalt; Selenium and Nickel
Inorganic Class 3 includes Lead; Chromium; Copper and Zinc

Source: Nilsson, K (1991), *World-wide trends in solid waste incineration*, Alliance for Beverage Cartons and the Environment, London

Table 1.2 Municipal waste incineration in selected developed economies

	Number of incineration plants	Quantity incinerated M/tonnes per year	Per cent of country's municipal waste incinerated	Energy recovery for heat and/or electricity
Sweden	23	1.8	55	All plants
Denmark	38	1.7	65	All plants
Germany	47	9.2	30	Most plants
Netherlands	12	2.8	40	Most plants
France	170	7.6	42	Most plants
Spain	22	0.7	6	Some plants
Italy	94	2.7	18	Some plants
UK	34	1.3	8	Few plants
USA	168	28.6	16	Most plants
Japan	1893	32.0	72	Few plants
Canada	17	1.7	19	Most plants

Source: Nilsson, K (1991), *World-wide trends in solid waste incineration*, Alliance for Beverage Cartons and the Environment, London

concentrated in Europe and encouraged by the increasingly tough national and EC level emission controls.[21] Proponents of incineration argue that emissions problems with dioxins can be technically solved and that the generation of electricity reduces the use of fossil fuels and therefore contributes towards a net reduction in the greenhouse effect.[22] Other research has claimed that incineration emissions are insignificant in relation to total air pollution in developed economies[23] and that toxins can be reduced by the separation of plastics from the rest of the waste stream by recycling activities.[24] The pro-incineration lobby have presented a range of arguments to illustrate the potential attractiveness of this option as an alternative to landfill for both the private sector waste management industry and local government for a number of reasons:

- No major alteration in the arrangements for waste collection is needed, and the management, construction and operation of incineration plants can be carried out in the private sector, thereby reducing the local and national tax burden.
- The volume of waste is reduced by as much as 90 per cent and may therefore extend available landfill space.
- There is a guaranteed market for the energy recovered from the plants, serving as a source of income for the recovery of capital costs in construction.
- The income from incineration rises over time, whereas the costs of landfill tend to increase, a trend likely to be reinforced with

the extension of civil liability to site operators for the future
environmental impact of sites.
* A further attraction for the use of incineration in urban areas is the
geographical concentration of waste in comparison with the
dispersed waste stream and cheaper landfill opportunities in rural
areas.

THE DEMUNICIPALIZATION OF WASTE MANAGEMENT

The rationale for extending the public sector role in waste management
since the late nineteenth century until at least the mid-1970s, rested largely on the perceived inadequacy of the private sector to carry out municipal
waste management in a cost-effective and environmentally acceptable
way.[25] Within developed economies, the emergence of a legislative framework for waste management can be traced to the public health fears
associated with rapid industrialization and urbanization in the nineteenth
century.[26] In the US, for example, the period from the 1890s to the outbreak of World War I saw a substantial shift away from the use of private
contractors and towards increased municipal control of street cleansing
and the collection and disposal of refuse. By World War I some 50 per
cent of American cities had a municipal collection system compared with
only 24 per cent in 1880.[27]

Figure 1.4 shows how expenditure on the collection and disposal of
municipal waste has risen sharply in the UK since the 1930s and this area
of local government services has emerged as a key target for the reduction
of public expenditure.[28] In the case of the US, cities spent some $300 mil-
lion on waste collection and disposal in 1940, rising to $1 billion by the
early 1960s. In 1980 the total had spiralled to exceed $4 billion, and among
local government expenditure was exceeded only by the costs of education and highways.[29] Since the late 1970s there has been a fundamental
shift in public policy away from the Keynesian model of high public
expenditure, towards an emphasis on the extension of market forces. This
has been accompanied by a change in the priorities of macro-economic
management towards the control of inflation and raising levels of profitability in the private sector. This has resulted in increasingly tight
controls on public expenditure, a decline of strategic planning, and cutbacks in social welfare programmes.[30] The changes during the 1980s have
been particularly significant in local government, where the role of municipal authorities has been substantially modified to reduce the national
and local tax burden on the 'wealth creating' private sector.[31]

The root of the contemporary political pressure for the privatization of
waste management can be traced to the mid-1970s, when the existing
structure of municipal waste management in the public sector, especially
in the US, began to be increasingly criticized on a number of grounds. The
growing environmental controls and technical complexity had left many
local municipalities in the operational role of waste collection only; the

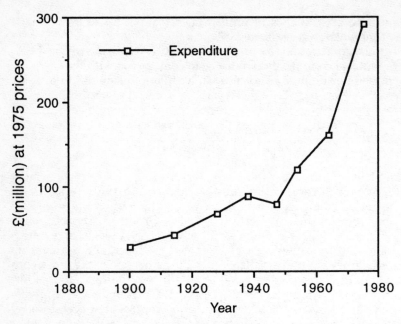

Figure 1.4 Expenditure on municipal waste collection and disposal
in the UK

Source: Dunleavy, P (1984) 'The limits to local government' in Boddy, M and
Fudge, C, *Local Socialism?*, MacMillan, London

spiralling costs of waste collection and waste disposal were a growing tax
burden; and the complexity of new developments in waste management
was exceeding available expertise in local government.[32] In the 1970s the
48 largest cities in the US were spending some 50 per cent of their environmental budgets on solid-waste management and a number of cities
began investigating whether there were alternatives to municipal service
provision.[33] A key concern was the high and rising costs of waste collection, which by the mid-1970s constituted over 70 per cent of the overall
costs of municipal waste management in the US, and this was attributed
by the political Right to labour practices in the public sector:

> A publicly operated collection agency tends to be less cost and efficiency
> conscious than a private one, especially if Solid Waste Management is fund-
> ed out of tax revenues. This usually results in excessively large crew sizes,
> abuse of sick leave, and the fostering of unionism to counteract the market
> power of the government.[34]

The capacity of the private sector to take on operational aspects of waste
management has developed substantially since the 1960s and 1970s. This
has been a result of the growing size and technological sophistication of
the waste management industry, with an increasing dominance of large

firms, including multinational companies specialized in the provision of municipal services. These companies are able to benefit from economies of scale and undertake research and development into profitable aspects of waste management such as landfill gas recovery and incineration with generation of electricity, functions that have become increasingly beyond the operational scope and financial capacity of local government. In the UK, for example, there has been a growing involvement of energy and water supply companies in municipal waste management in response to a number of developments: the privatization of public utilities; the creation of market opportunities for the provision of local government services through competitive tendering legislation; and the decline of other less profitable sectors of technical engineering such as nuclear energy since the 1970s.[35] The economic recession of the 1990s has given further impetus to this process as local government struggles to maintain services and reduce costs, in the face of the private sector lowering the cost of tendering bids, in order to increase its market share.[36] In the case of London the wide-ranging strategic local government role in waste management has been radically diminished since the mid-1980s to a fragmentary panoply of client authorities with narrowly defined fiduciary responsibilities and negligible input into policy, planning and research. In Hamburg and New York, by contrast, waste management remains a municipal service but faces severe financial pressures and major policy dilemmas over the appropriate role for the private sector in the construction and operation of, for example, incineration plants.

The growing cost and technological sophistication of waste management in the UK has already begun a major restructuring of the private sector as smaller firms are 'squeezed out' by their larger competitors. The increased operational costs are forcing a rationalization of the highly fragmented waste management industry, which has over 4000 companies (on some estimates) and it is predicted that the large industry shakeout in the waste management industry will be accompanied by a wave of acquisitions, similar to those that transformed the US waste industry in the 1970s when environmental standards were raised.[37] As a result of the changing context for waste management, the industry has been undergoing a diversification and internationalization of its activities:

> The waste management industry is going through the greatest transformation in its history as a result of the large increase in public concern for the state of the environment.... . Governments have been spurred to take action and the trend is towards stiffer national regulations for the control of all wastes... . The growth in the amount of waste being handled and the increasing sophistication of the techniques being used means that waste management is very big business.[38]

2

Recycling in perspective

Estimates of the maximum achievable level of recycling for municipal waste range up to as much as 90 per cent by weight though current levels of recycling lie far below technically achievable levels.[1] By the late 1980s 12 communities in the US had recorded recycling rates of over 25 per cent, and by 1989 Seattle had reached a rate of 37 per cent – the highest US figure for a larger city. Similarly the German city of Heidelberg has recorded a recycling rate of 37 per cent accomplished through a combination of a dual bin system for putrescible kitchen and garden wastes accompanied by an extensive network of on-street collection facilities for glass and other materials.[2] It is important to note that almost all waste management systems achieving recycling rates in excess of around 20 per cent not only make use of kerbside collection systems but also involve the composting of a significant fraction of putrescible wastes with implications for the cost of these programmes in comparison with disposal of waste by landfill or incineration.

THE DIVERSE RATIONALE FOR RECYCLING

The wide-ranging contemporary rationale for recycling in developed economies mirrors the diversity of environmentalist thought and ranges from attempts to reduce the costs of waste management to the promotion of alternative utopian conceptions of society. Although there is a political consensus over the need to promote recycling this does not extend to the precise objectives of policy or over the most appropriate means of promoting recycling and sustainable waste management strategies.

There are a variety of environmental justifications for recycling which can be found in the literature:

- the conservation of finite resources as a move towards 'steady state' or sustainable economies;[3]
- the reduction in energy consumption and more recently, the limiting of greenhouse gas emissions;[4]
- the control of pollution involved in the production process, extractive industries, and in the final disposal of waste;[5] and
- the environmental education benefits of participation in recycling.[6]

If the analysis of the environmental impact of recycling is extended throughout the whole cycle of production and consumption a complex picture emerges as illustrated in Figure 2.1. This has been the subject of a number of so-called ecobalance studies to compare the environmental impact of different kinds of packaging materials.[7] The recycling of different components from the waste stream has a differential environmental impact depending on which materials form the focus of recycling policy, as illustrated in Table 2.1, which compares the pollution associated with

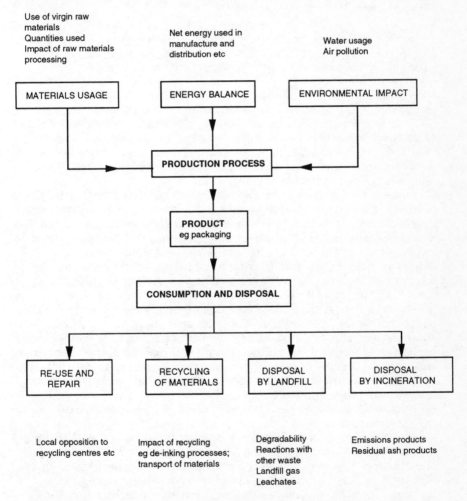

Figure 2.1 The environmental impact of different components of the waste stream

Source: Adapted from Gehrke, C (1988), Möglichkeiten und Grenzen des abfallarmen Einkaufs', in Institut für ökologisches Recycling, *Abfall Vermeiden*, Fischer Taschenbuch Verlag, Frankfurt am Main.

Table 2.1 Environmental benefits derived from substituting secondary materials for virgin resources

Per cent reduction	Aluminium	Steel	Paper	Glass
Energy usage	90–97	47–74	23–74	4–32
Air pollution	95	85	74	20
Water pollution	97	76	35	–
Mining wastes	–	97	–	80
Water usage	–	40	58	50

Source: Letcher, R C and Shiel, M (1986), 'Source Separation and Citizen Recycling', in Robinson, W D *The Solid Waste Handbook: A Practical Guide*, John Wiley, New York

the processing of primary and secondary raw materials. Note, for example, that the recycling of materials such as aluminium on a tonne for tonne basis is far more environmentally significant than glass because of the impact of bauxite extraction on areas of high biodiversity such as tropical rainforests and the massive inputs of energy required in processing of aluminium from bauxite ore.

There is evidence that the promotion of materials recovery may not necessarily be environmentally preferable to energy recovery in all cases.[8] As a recent illustration, research by the International Institute for Applied Systems Analysis in Austria has suggested that the maximum level of paper recycling leads to greater emissions of sulphur dioxide, nitrous oxides and carbon dioxide in comparison with a component of energy recovery by incineration allowing a reduction in the overall use of fossil fuels.[9] In contrast a US study found that recycling conserves 3-5 times as much energy as can be generated from an equivalent quantity of waste being incinerated.[10] The key issue with the use of various Life Cycle Analysis and ecobalance type studies is that the selection and omission of particular factors as depicted in Table 2.2 and the methods of calculation adopted will inevitably make the results reflect previously held assumptions and that different parts of the packaging and waste management industry are busy commissioning studies to promote their own particular product or technology as environmentally and economically superior to their competitors.

An important economic benefit claimed for recycling is that it may reduce the costs of waste disposal for urban areas which face fewer cheap landfill opportunities. As a typical illustration, John Young of the Washington based Worldwatch Institute argues that 'recycling and composting offer a cheaper, more effective alternative to incineration, one that can cut landfill needs to a minimum'.[11] Yet the experience of London, Hamburg and New York suggests that this is unlikely to be the case because the costs of comprehensive materials recycling programmes have been consistently underestimated in relation to more profitable alternatives for waste disposal such as incineration with energy recovery, which has important implications for the promotion of market-based recycling and waste reduction strategies.

Table 2.2 The universal production matrix

9. Waste disposal	8. Transport	7. Consumption	6. Retail and marketing	5. Production	4. Transport	3. Raw materials processing	2. Transport	1. Extractive industries		
									Energy usage	Environmental impact
									Raw materials usage	
									Land usage	
									Water usage	
									Water quality	
									Waste production	
									Air pollution	
									Ground contamination	
									Pollution effluents	
									Radioactive waste	
									Flora	
									Fauna	
									Impact on countryside and green space	
									Work satisfaction	Social impact
									Health and safety at work	
									Participation in decision making	
									Cultural pluralism	
									Leisure time	
									Economic costs	Economic impact
									Macro economic effects	
									Implications for the world economy	
									North–south global dimension	
									Wealth creation and economic activity	

Source: Adapted from Teichert, V (1989), 'Die Produktlinienanalyse: Möglichkeiten für ihre politische Implementation', paper presented to the conference Ökologische Abfallwirtschaft, 30 November to 2 December, Technische Universität, Berlin.

In the period since the 1960s, a range of economic and social justifications for recycling have been made: the potential for reducing the balance of payments deficit in raw materials;[12] the need for increased geo-political resource security against producer cartels and the danger of resource wars over oil and other vital commodities;[13] and the need to redress the imbalance in global resource use to allow the economic development of poorer countries without jeopardizing efforts to protect the global environment.[14] A further justification for recycling is the potential for job creation, especially in depressed areas with high unemployment. Examples include the advocacy of community based industries for paper and glass making[15] and also the renovation and repair of household goods at recycling centres.[16] This job creation aim has also formed part of most radical alternative waste management strategies and is a key element of the so-called environmental Keynesianism of the mid-1980s where expenditure on environmental regeneration is seen as a means to stimulate local economies suffering from the structural decline and relocation of existing sources of employment.[17]

Recycling is often portrayed within environmentalist literature as integral to the conception of an alternative society founded on small-scale community based industries and commune type human settlements. The aim is to create self-sufficient sustainable urban communities based around local production for local consumption from local resources.[18] These ideas are linked with utopian socialist and anarchist strands of environmentalism but the implicit emphasis on rural communes does not provide a useful blueprint for tackling the contemporary waste management crisis facing urban societies. Community based recycling and waste reduction have also been conceived as a practical example of a 'soft' technological path for waste management and the distinction between technocentric 'hard' and ecocentric 'soft' technology has been framed in terms of the choice between incineration technologies for energy recovery and community scale recycling initiatives focused on materials recovery.[19] Some writers have sought to show that the distinction between 'soft' and 'hard' approaches to waste management is paralleled by the wider distinction between ecocentric and technocentric approaches to the environmental crisis,[20] but since the 1980s the clarity of this distinction has been weakened by the rise of 'green consumerism' and the emerging political consensus over environmental sustainability.[21]

THE HIERARCHY OF RECYCLING OPTIONS

Although the contemporary emphasis of recycling policy is on materials in household waste, it is useful to picture recycling as a hierarchy of potential options ranging from waste reduction at source to energy recovery from landfill or incineration. Note that each of the steps illustrated in Figure 2.2 varies in terms of its potential environmental impact, both locally and globally; the levels of public participation required; the potential costs involved; and the degree of technological sophistication. Most critically, each step implies a different balance between the role of the state and the market, such that maximum reduction of waste at source would involve widespread intervention in the economy whereas energy recovery can be pursued with minimal state intervention as a profitable venture for the private sector.

- **Waste reduction**
  This is the first step within the hierarchy of options. Waste minimization and waste prevention in the production process is seen as integral to radical environmentalist waste management strategies. For example, the Berlin based Institute for Ecological Recycling claim that up to 90 per cent of pollution results from the production process, not in the eventual disposal of products, and this underlies the rationale of the radical environmentalist demands for a focus on waste reduction at source rather than a policy emphasis on materials in post-consumer waste.[22]
- **Pre-consumer recycling**
  Where the production of wastes is unavoidable the second step is their re-use within the production process itself. This is easier than post-consumer recycling since uncontaminated and economically handleable quantities of waste are in proximity to industry as potential new raw materials. Some items sold as recycled products are made in just such a way, and contain no post-consumer waste, whereas in other cases there may be a mixture of pre- and post-consumer wastes which have been recycled. The economic and logistical advantages of recycling within the production process are reflected in the higher rates of industrial recycling in comparison with the recycling of similar materials from the municipal waste stream.[23] Some environmentalists have assumed – mistakenly – that the economic advantages of recycling industrial wastes are paralleled in the recycling of municipal wastes;[24] but the dispersed and mixed nature of municipal wastes raises a range of economic, technical and logistical complexities for the recovery of re-usable materials.
- **Product re-use**
  The third step is the re-use and repair of products to prolong their usefulness before entering the waste stream. This includes the use of returnable beverage containers and the elimination of built-in obsolescence in consumer durables. The re-use and repair of products has also been linked with community based recycling projects and also with the generation of employment.
- **Primary recovery**
  The fourth step is the recycling of the waste stream to create new raw materials. This forms the main focus of this book and of contemporary recycling policies within developed economies. It includes a variety of measures such as the use of on-street collection facilities for paper and glass; the magnetic separation of ferrous scrap at some

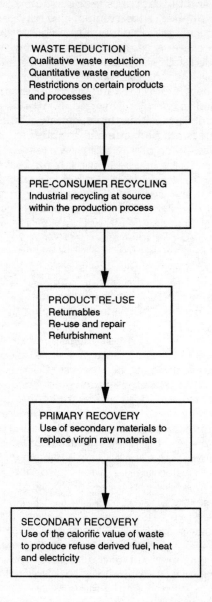

Figure 2.2 The hierarchy of recycling options

Source: Gandy, M (1993), *Recycling and Waste: an exploration of contemporary environmental policy*, Avebury, Aldershot.

incineration plants and waste transfer stations; kerbside recycling programmes and the production of compost from putrescible waste.

- **Secondary recovery**
The fifth step is the recovery of energy from waste. This includes the production of refuse derived fuel; the use of incineration plants integrated into combined heat and power systems; the recovery of landfill gas as a fuel source; and high temperature pyrolysis of tyres and plastics. The energy recovery option of incineration remains highly contentious in terms of its legitimacy as a form of recycling because there is no challenge to the underlying dynamic of a growing waste stream.

THE RECYCLING OF POST-CONSUMER WASTE

The central focus of contemporary recycling policy in developed economies is on the fourth step within the recycling hierarchy – the recovery of re-useable materials from the municipal waste stream. There are three main methods of recovering materials from the waste stream:

1. 'collect' or kerbside schemes;
2. 'bring' systems based around the use of on-street collection facilities and recycling centres; and
3. centralized sorting facilities, sometimes linked with energy recovery from the calorific value of waste.

Table 2.3 shows that these three organizational approaches differ in the levels of public participation required; the capital and operating costs; the technical and logistical difficulties in their implementation; and the potential levels of materials recovery which can be attained.

The use of 'bring' systems and on-street collection facilities

With the operation of 'bring' systems, individual citizens take recyclable materials to a public collection site. The facilities can vary from single material bottle banks to multi-material recycling centres handling the whole range of recyclable materials, renovating furniture and other consumer durables and offering specialist advice on the disposal of household toxic waste. On-street collection facilities such as bottle banks and recycling centres were introduced from the late 1970s onwards in developed economies in response to the perceived difficulties with the existing approaches to recycling. In the UK, for example, the 1970s saw the decline of public sector kerbside paper collection schemes, because of their increasing labour costs and the weakness of the secondary materials market for waste paper.[25] The advantages of on-street collection facilities are

Table 2.3 A comparison of different organizational approaches to
materials recycling

Method of recovery	Advantages	Disadvantages
Collect systems, eg Kerbside collection Multiple wheeled bin systems Dual bin systems for 'wet' and 'dry' wastes	• High levels of materials recovery • Wide range of materials can be recovered, including putrescible wastes and toxics • Low levels of materials contamination • High levels of public participation	• High capital costs • High labour costs • Reliant on continuous public cooperation • Difficult in congested urban areas
Bring systems, eg On-street collection facilities for glass recycling centres and civic amenity sites	• Low labour costs • Low capital costs • Local government can take a purely enabling role	• Low levels of materials recovery • Local environmental impact • Difficult in congested urban areas • Difficult to provide high density of facilities in areas of low population density • Contamination of materials
Centralized sorting plants, eg Transfer stations with mechanical separation technologies or manual sorting	• Low labour costs • Can be combined with the production of refuse derived fuel or energy recovery technologies • No changes needed in the arrangements for waste collection	• High capital costs • Political opposition to plants, especially when combined with incineration • Moderate levels of materials recovery but problems of contamination • Difficult for small municipal authorities to fund and operate without reliance on the private sector • May require hand sorting of mixed waste

Source: Gandy, M (1993), *Recycling and Waste: an exploration of contemporary
environmental policy*, Avebury, Aldershot.

mainly economic and political since their provision and operation are cheaper than kerbside collections, in terms of both capital and labour costs. The role of local government can be merely an enabler for the activities of the private sector and the voluntary sector. Both local government and central government can therefore claim that they are responding to environmental concern, but with a minimal increase in public expenditure.[26] Yet the disadvantages of on-street collection facilities are mainly environmental: the level of recycling is lower than with the use of kerbside 'collect' systems, both in the quality and quantity of collected materials; the local environmental impact of facilities such as bottle banks may outweigh the wider environmental benefits and undermine public confidence in recycling as a long-term component of municipal waste management; and the patterns of usage involve the widespread deployment of cars to transport glass, which is among the least important materials for recycling in terms of its environmental impact, as indicated by Table 2.1.

The use of comprehensive kerbside collection and dual-bin systems

Under contemporary 'collect' systems, households separate the materials, which are then collected by the public sector, private contractors, or the voluntary groups, sometimes in combination with each other. Historically, many municipal authorities ran their own source separation programmes for household waste but the post-war period has seen a marked decline in sophisticated waste collection services provided by local government. This can be attributed largely to the escalating labour costs of such programmes in comparison with the routine collection and disposal of waste. However, a 'collect' system can form part of a more sophisticated routine waste collection service, such as the use of multiple bin systems. In Munich, for example, there is a dual bin system in operation where putrescible *biomüll* for composting is collected separately from the remaining *restmüll*.[27] In Denmark there has been the use of a three bin system for mixed recyclables (paper, glass and metals); putrescible waste for composting; and the remainder for disposal by landfill or incineration.[28] If the recyclables are collected separately from routine waste collections as in the 'blue box' or 'green bin' schemes, this normally involves higher costs than combined collection systems. In the case of the 'blue box' scheme, mixed recyclables are taken to a centralized sorting facility, and some ten million households are now participating in blue box schemes in North America.[29]

The established advantages of kerbside collection systems are the higher recycling rates which are achievable in comparison with 'bring' systems and their popularity with the public.[30] In contrast to 'bring' systems the main problems with the use of kerbside collection systems are the higher capital and labour costs for local government. A survey carried out by Warren Spring Laboratory for the UK government found that blue box kerbside schemes cost around £85 to £175 per tonne in comparison with around £16 to £36 for bring systems.[31] These higher costs can be reduced only through extensive private sector sponsorship, state subsidy or voluntary sector involvement. In Adur, West Sussex, the UK government's target of 25 per cent recycling by the year 2000 is reported for a kerbside collection scheme with financial backing from the Brussels based European Recovery and Recycling Association (ERRA), which represents the interests of 27 companies including British Steel, Nestlé, Perrier and Pepsi-Cola International. An examination of the funding for private sector sponsored kerbside collection schemes suggests that they are sophisticated lobbying campaigns to counter any legislative controls on the private sector packaging industries. The private sector will only participate in these loss making schemes as part of a dual strategy to influence legislative developments for waste management and also maintain or improve their market share by demonstrating their 'green' credentials. If kerbside collection schemes are not sponsored as part of the lobbying activities of the private sector or financed by central government the costs will be borne by the local population through higher charges for their routine waste management services.[32]

Centralized sorting and capital intensive resource recovery

The use of centralized sorting rather than individual citizens as the main means of separating re-usable materials from the waste stream has had two main phases of use. The first phase, until the early 1960s, formed part of the operational aspects of waste incineration, and involved the recovery of a range of materials, usually by hand picking from a moving belt. The practice of hand picking has declined in developed economies because of the high labour costs and the difficulties in recruiting and retaining labour for unpleasant manual tasks, though hand picking is still used in a number of sorting plants for mixed recyclable materials as in Harlem's materials recovery facility serving New York City.[33] The second phase, during the 1970s, involved the increased use of mechanical sorting of the waste stream, often with the production of refuse-derived fuel. The promotion of centralized sorting plants and the production of refuse-derived fuel has declined steadily since the 1980s as a result of technical and economic difficulties combined with the advent of cheaper methods of waste separation by the use of on-street collection facilities, which have only been widely introduced since the late 1970s. The interest in refuse-derived fuel in the 1970s has also waned because the predicted shortages and price rises for primary sources of energy have not materialized, and environmental concern has shifted from the 'energy crisis' of the 1970s to a new focus on global climatic change caused by the emission of atmospheric pollutants. Since the 1980s the emphasis on centralized and capital intensive recycling and waste management has been moving towards the profitable recovery of the calorific value of waste to produce electricity in the new generation of incineration plants.

THE CONTROL OF PACKAGING WASTE

Packaging contributes about 35 per cent of the weight and 50 per cent of the volume of household waste in developed economies and the proportion of packaging within municipal waste continues to grow steadily.[34] However, the factors affecting the growth of packaging lie largely outside the control of local and regional governments who are responsible for managing the municipal waste stream, and the packaging policy debate is increasingly focused on national and international policy towards the packaging, retail and marketing sectors of the economy, as illustrated by the German Packaging Ordinance introduced in 1991. In the UK alone, the packaging industry is estimated to be worth at least £5.3 billion a year with over 75,000 employees[35] and the political strength of the packaging lobby is reflected in a range of influential organizations such as the Alliance for Beverage Cartons and the Environment, the Industry Council on Packaging and the Environment, and the European Recycling and Recovery Association.

The growth of packaging waste over the post-war period is an outcome of a major transformation in both economy and society within developed economies, associated with rising levels of private consumption for mass produced commodities.[36] The historian Martin Melosi describes the changes in the US:

> The dramatic rise in the automobile, chemical and electrical industries during the 1920s superimposed a consumer-oriented, service-oriented economy over the nation's industrial–agricultural base. The most obvious impact of this change, with respect to the refuse problem, was the appearance of numerous and diverse products and packaging materials – such as paper goods, plastics, toxic chemicals and synthetics – which would ultimately find their way into American trash heaps.[37]

Trends within retailing over the post-war period associated with the growth of supermarkets and the rise of mass consumption have led to the increased use of smaller one-way containers for food and beverages, which formerly would often have been sold in larger returnable containers or, in the case of many bakeries and greengrocers, with minimal packaging.[38] Over 80 per cent of packaging waste in developed economies is now calculated to be derived from food and beverages[39] and the proportion of packaging-derived municipal waste continues to grow, adding substantial costs to many essential products such as food.[40]

Since the late 1960s concern has focused not only on the rise in the quantity and variety of packaging over the post-war period but also on the environmental impact of packaging and the increasing proportion of materials which are difficult to recycle, such as mixed plastics and laminated cartons. As a result, there have been calls from within the environmental movement for state controls on superfluous and non-returnable packaging.[41] In the case of plastics, production and use in the OECD nations has more than quadrupled since the late 1960s and world-wide production is doubling every twelve years.[42] Plastics represent the fastest growing portion of municipal waste, increasing from just 1 per cent of the US waste stream by weight in 1960 to 2.7 per cent in 1970 and 7.2 per cent by 1984. This is equivalent to a 720 per cent rise in less than 25 years and it is predicted that plastics will account for almost 10 per cent of the municipal waste stream by the year 2000.[43] This rapid rise in the use of plastics reflects their unique advantages in terms of their chemical and physical properties, especially for packaging, and plastics manufacturers are now targeting metal cans as their next major market opportunity for expansion.[44]

Plastics are produced primarily from non-renewable oil and gas, and are replacing materials which can be much more easily recycled. In comparison with glass and metals, for example, plastics change physically and chemically during reprocessing. This means that the number of times plastic can be recycled is very limited, as most polymers cannot be recycled back into the original type of packaging. Additional difficulties are that some polymers may absorb small quantities of materials they contain, such as pesticides, and that plastics cannot be satisfactorily sterilized. These characteristics limit the potential for recycled plastic food packaging.[45] Research has suggested that the production of plastics involves four times the energy used in glass manufacture, and that there are greater pollution emissions than in paper or glass production. There are serious emissions concerns over the recovery of the chemical constituents of plastics by pyrolysis or the energy value through incineration.[46] The US Environmental Protection Agency estimates that 71 per cent of lead emissions and 88 per cent of cadmium emissions from the incineration of municipal solid waste are derived from plastics packaging materials.[47] As a result of these difficulties, many environmentalists have argued for the reduced production of plastics, and the banning of potentially toxic polymers such as PVC (polyvinyl chloride).[48] PVC is the most hazardous type of plastic in common use, since the presence of chlorine means that organochlorines are created both in the production process and during final disposal by incineration. Organochlorines are a family of chemicals including DDT, PCBs, dioxins and furans, some of the most toxic chemicals known.[49]

In contrast, the European Plastics Manufacturers Association (APME) comprising most of European plastics manufacturers, has undertaken Life Cycle Analysis of different plastics packaging. Results suggest that PET (Polyethylene terepthalate) containers have a lower environmental impact than paper or glass and the organization has been seeking to influence the EC Packaging Directive in order to protect the plastics packaging industry.[50] The plastics industry clearly take the prospect of legislative control very seriously, and have countered the threat through the promotion of biodegradable polymers and the labelling of a variety of polymers as recyclable (irrespective of the available markets and recycling infrastructure). Most plastics recycling is, however, best referred to as 'linear recycling' where polymers such as PET and HDPE are converted into inferior products such as concrete and wood substitutes, thereby replacing other, more recyclable materials and not affecting the use of virgin materials in plastics production for packaging.

Furthermore, much of the plastics supposedly recycled are being shipped to developing countries where there are more lax environmental regulations, limited community opposition to waste processing facilities and lower labour costs. A Greenpeace survey of plastics exports from Canada to Asia in 1992 found that between 25 and 40 per cent of the imported wastes are dumped and that the recycling processes emit noxious fumes posing a threat to the workforce:

> None of the facilities visited were equipped with ventilation systems or respiratory protectors for the workers. The men, women and children working at these machines were constantly inhaling the fumes in work conditions that would never be allowed in Canada.[51]

Concern with the growing strength of the packaging lobby in Europe has led to the formation of the European Sustainable Packaging Network involving some 55 different non-governmental organizations such as Greenpeace and the Women's Environmental Network. They have called for the elimination of superfluous packaging, a substantial reduction in one-way laminated packaging and composites and a shift of emphasis away from materials recycling towards the use of returnable packaging.[52] A recent report by Netherlands Friends of the Earth into packaging in Eastern Europe has revealed how multilateral development banks are favouring the one-way packaging industry such as the laminated carton producer Tetra Pak and that in Hungary alone, the market share of disposable laminated fruit juice and milk cartons has increased from 10 per cent to 60 per cent in the period 1988–1991 alone.[53]

The argument for returnable containers is based on the growing contribution of one-way containers to the volume of waste. One study suggests that the proliferation of smaller one-way containers has involved a four-fold increase in waste compared with the use of larger returnable containers, and that on a one-for-one basis, returnables handle, on average, forty times as much volume of products as one-way containers.[54] The positive impact of returnables is dependent on a sufficient number of cycles between consumers, retailers and producers.[55] Technical developments in returnable packaging during the 1980s have lowered the number of cycles needed in order to have a positive environmental impact.[56] In terms of the economic impact, it is maintained that job losses in the one-way packaging industry would be more than offset by employment created within the recycling and waste management sector. An example is the creation of community based industries for the reclamation, renovation and repair of goods.[57]

In contrast, the packaging lobby has consistently claimed that packaging provides benefits in terms of safety, enhanced quality and convenience, and in particular, reduced food spoilage.[58] It is contended that advances in packaging technology have made it lighter and increasingly efficient in terms of both materials and energy usage,[59] and that with the partial replacement of glass by plastics, the weight of packaging in municipal waste has declined in recent years.[60] As an illustration, the average weight of a one pint milk

bottle in the UK has fallen from 538 grams in 1940 to 245 grams in 1990.[61]
Over longer distances, returnables are held to have no superiority in either
economic or environmental terms, and as a result of glass recycling it is
claimed that the amount of waste from disposable glass beverage containers is less than in 1974.[62] One German study has estimated that with a
complete switch to glass containers, there would be a rise in the weight of
municipal waste by 40 per cent and an increase in volume of 256 per cent.
The use of energy would grow by 201 per cent and the cost of packaging
would climb by 212 per cent.[63]

The US experience has been an important source of opposition to the
introduction of mandatory deposits or controls on packaging in Europe.
In the early 1980s, nine states introduced a system of mandatory deposits
and charges on beverage containers, which were shown by some studies
to have far-reaching and largely unforeseen consequences:

- The price of beverages rose by an average of 28 per cent while sales
  fell by an average of 33 per cent, because the handling expenses for
  the forced deposit system could not be covered by the scrap value of
  the returnable packaging.
- Fewer brands of drinks became available.
- There were job losses in the glass industry, creating unskilled jobs at
  the expense of skilled jobs.
- There was an increasing share of aluminium and plastics
  (particularly PET) in the packaging market.
- The level of non-redeemed deposits rose as high as 50 per cent in
  some states.
- There was a negligible impact on litter.
- Many established voluntary sector recycling schemes disappeared.
- The level of recycling was not markedly better than in neighbouring
  states using voluntary recycling schemes for beverage containers.[64]

Yet these experiences are in stark contrast to the findings of the 1989
Moreland Commission investigation into the operation of the returnable
container legislation in New York State, where a range of social and environmental benefits were identified (see Chapter 4).[65]

In summary, the problem of packaging-derived waste raises three key
issues in waste management:

1. The practical difficulties in influencing the size and composition of
   the municipal waste stream.
2. The ideological debate surrounding the degree and nature of state
   intervention in the production process, retail and marketing.
3. The complexities of the production and consumption pattern involved in the creation and disposal of waste and the difficulties in evaluating the environmental impact of alternative kinds of packaging.[66]

PUBLIC PARTICIPATION AND RECYCLING

Public participation is essential for 'bring' and 'collect' systems which rely on the sorting of waste by households. The increased role of public participation in recycling reflects the growing emphasis of environmental policy on individual responsibility and the 'green consumerism' facilitated through eco-labelling schemes and public awareness campaigns. Previous peaks in public involvement in recycling in developed economies occurred during wartime as part of the national war effort.[67] In the London Borough of Enfield, for example, there were collection points on every street corner where six different waste fractions were left by householders: kitchen waste (without tea leaves) for pig swill, old bottles, cans, bones, rags and newspapers.[68] Similar patterns of public participation for reasons of economic autarchy even pre-dated the war effort in Germany and parallel more recent recycling efforts in the former Communist states to buffer them from the need to purchase essential raw materials on world markets.[69] Over the post-war period the recycling of waste for economic survival has only been commonplace in developed economies during times of economic austerity and hardship[70] and is now most widespread in larger Third World cities where many thousands of people survive by salvaging reusable materials.[71]

There have been a number of studies by behavioural psychologists to examine the motivation of participants in recycling schemes and the effects of different kinds of inducements on the rate of participation. A study of recyclers in Michigan found that the personal satisfaction associated with environmentally responsible behaviour contributed to citizens' sense of well-being and that conservation behaviour might be carried out without the need for any form of financial inducement.[72] There is also evidence that participation will rise if the scheme donates its proceeds to a charitable organization. In San Francisco, for example, a fourfold increase in usage of on-street collection facilities and recycling centres was reported when the proceeds were given to AIDS charities.[73] In contrast, other studies suggest that financial inducements are always necessary to gain the participation of less environmentally motivated 'non-recyclers'.[74] Incentives can take a number of forms, ranging from direct charges for waste management services to systems of fines and penalties under mandatory recycling programmes, as in New York.

Studies have suggested that shared bins, such as chutes with paladins for high rise housing, present a psychological barrier to effective participation because of their anonymity.[75] The relative importance of different features associated with shared bins or paladins is unclear, since their use indicates both the physical characteristics of dwellings (such as lack of space and the presence of common areas) and also socio-economic features such as higher levels of deprivation in much inner urban high rise housing. For example, a study of the 'green bin' system in Germany found that the high levels of materials contamination in high rise housing areas could be attributed to a combination of five main factors: the housing structure and degree of anonymity; the social structure and levels of deprivation; the degree of environmental awareness among householders; the extent of public information on the scheme; and the frequency of collections.[76]

The use of direct charges to citizens for their waste collection service based on the volume or weight of waste collected has been used by some municipalities in the Netherlands, Germany, New Zealand and the US. For example, in the town of Hoofddorp in the Netherlands and Billingheim in southern Germany, dustbins have even been fitted with a special measuring device for this purpose.[77] This is a clear example of the 'polluter pays principle' applied directly to individual households, yet there are difficulties with the use of 'end of pipe' direct charging for consumer waste. It is also doubtful whether households should be held financially responsible for waste derived from unnecessary packaging or toxic materials. The use of direct charges for essential municipal services is socially regressive because basic household necessities constitute a larger proportion of the budget of low income households.[78]

THE SECONDARY MATERIALS MARKET

There are a number of factors behind the contemporary weakness of the secondary materials market:

- The mis-match between the supply of recycled materials and demand for recycled products, which has followed increased recycling efforts for some materials.
- The competition with virgin raw materials, which often enjoy extensive state subsidies.
- The difficulties of ensuring consistent supply of uncontaminated materials in economically handleable quantities.
- The current impact of the world recession on levels of economic activity and the demand for raw materials.

Evidence suggests that recycling has been hampered by the secondary materials market for many decades. An important study published by the US Environmental Protection Agency in 1972 showed that the level of recycling in the US had progressively declined over the post-war period. This decline was attributed to the difficulties faced by the secondary materials industry, especially for paper, textiles and rubber. This was dominated by small-scale, often family run enterprises, who found it increasingly difficult to compete with virgin materials because of rapid technological developments in the capital intensive extractive industries for the processing of raw materials. It was increasingly difficult and expensive to collect recyclables from householders, for whom sorting wastes had become a repugnant activity. Meanwhile, virgin materials had improved steadily in quality and were increasingly competitive in price. Furthermore, the waste stream now included increasing quantities of materials which were difficult to reclaim, such as plastics and composite items, for which no secondary materials market existed.[79] The US paper recycling rate fell from 35 per cent in 1944 to around 17 per cent by 1973 and in the case of rubber there was a fall from 19 per cent recycling in 1958 to only 8.8 per cent in 1969. Similar patterns of decline were also noted for aluminium, copper, lead, and ferrous scrap.[80] Processes of decline within the secondary materials industries have also been described in the UK. In the case of tyres, the only major market for recycled rubber, that of carpet underlay, collapsed in the 1960s and early 1970s with the advent of plastic foam-backed carpets.[81] Another example is textiles recycling. Although long established in the UK, demand for recycled fabric and fibre waste has fallen over the last 20 to 30 years and many textile reclaimers have been forced to cease trading. Post-consumer textile waste now comprises a complex mixture of natural and synthetic fibres making the identification, sorting and grading of textile waste for different end uses an increasingly expensive labour intensive operation.[82]

The most important recent example of a price collapse is the glut of low grade waste paper which currently prevails in Europe and North America, and has driven down prices as shown in Figure 2.3. In the UK, the swamping of the market for low grade paper has seen prices fall rapidly in the late 1980s and early 1990s and some ten per cent of the UK paper recycling industry went into liquidation during 1992.[83] This has led many paper recovery schemes into bankruptcy and meant that quantities of waste paper collected for recycling by the general public have been dumped at landfill sites.[84] Several factors have contributed to the current weakness in the waste paper market: the increased participation in recycling by the public during the current wave of environmental concern; the attempts by Germany and the US to mandatorily increase the recycling of paper, thereby swamping the world market; the dearth of capacity in the de-inking and paper reclamation industry; the competition with high quality cheap virgin pulp imports; and the high demands that merchants and mills have been stipulating for quality of waste paper in the face of uncertainty over the marketing of recycled paper products.[85] The state of the waste paper market has deteriorated still further in the 1990s as a result of world recession lowering the demand for paper products and price cuts for virgin pulp in an attempt to stimulate demand by the depressed paper industry now facing a spate of bankruptcies in North America. The current recession in Germany has led to plummeting demand for waste paper and desperate suppliers are looking for new markets elsewhere in Europe, contributing to a fall in waste paper prices in the UK alone of 16 per cent between May 1992 and May 1993.[86]

The weakness of the secondary materials market has been a key barrier to higher levels of materials recycling throughout the post-war period and earlier, and has had three main impacts on recycling activity:

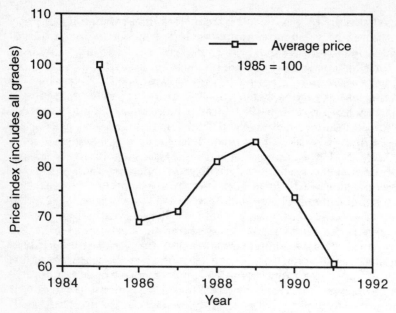

Figure 2.3 The price of waste paper and board in the UK, 1985–91

Source: UK Central Statistical Office, Cardiff

1. The periodic collapse of markets in response to increases in the amount of collected materials during periods of environmental concern.
2. The long-term decline in some sectors, such as rubber and textiles, in response to changing patterns of production and consumption and advances in raw materials processing technology.
3. The innate weakness of the secondary materials market, reflected in the impact of economic recession on the demand for waste paper and other commodities.

These difficulties are scarcely addressed by most programmes to radically increase the level of recycling towards technically achievable levels. The problem is merely compounded by more effective collection systems in the absence of an intricately coordinated and planned attempt to influence the whole cycle of production and consumption. The contemporary vicious circle is a result of increasing quantities of materials collected weakening the market, leading to falling prices and the disruption of recycling schemes. This, in turn, serves to undermine public confidence in the promotion of comprehensive recycling policies, and has indirectly served to increase attention on alternatives to materials recycling such as harnessing energy from waste.[87]

THE ECONOMIC VIABILITY OF RECYCLING

A consideration of the economic aspects of recycling is important because a key reason for the promotion of recycling is frequently argued to be the potential of schemes to generate a profit from the sales of materials and the creation of savings in waste disposal costs. The recycling of some materials such as glass via the 'bring' system may be economically viable in some cases[88] but most evidence suggests that comprehensive recycling programmes where kerbside 'collect' systems are extended to paper, putrescibles, plastics and other materials are more expensive than routine waste collection and disposal by landfill or incineration.[89]

Many municipalities have seen a rise in their income from sales of materials since the early 1980s, reflecting a greater variety and larger quantity of materials recovered and sold, but the range of associated overheads and other costs has also risen. The costs and overheads render the calculation of recycling expenditure much more difficult than that of income derived from the sales of materials. A key question concerns the distribution of costs, such as labour and capital equipment, between the recycling budget and the municipal waste management budget as a whole.[90] This difficulty in determining the net income from recycling is important, because this informs the rationale for further expanding the policy on purely economic grounds, whether this be in terms of reducing the costs of waste disposal or the promotion of recycling to make a profit. If it can be demonstrated that a recycling scheme has resulted in a net income then a rationale for the policy becomes an additional source of revenue for the municipal authority as is claimed by some London Boroughs.[91]

An important theme developed in this book is that within the contemporary legislative and administrative framework for municipal waste management, comprehensive recycling programmes are more expensive than waste disposal by landfill or incineration. The relative costs of recycling in relation to waste disposal may shift in response to the evolving legislative framework for waste management, but significant changes arising from the use of input taxes on virgin raw materials or mandatory controls on consumer products will be the outcome of a fundamental change in the contemporary political context for environmental policy making.

RECYCLING AND 'MARKET FAILURE'

The last decade has seen increasing interest in the use of economic or market-based policy instruments (MBIs) for environmental protection rather than the regulatory role of government on the premise that not only are MBIs cheaper but that they are a more effective means of protecting the environment.[92] The market-based approach to environmental policy is based on the belief that a properly functioning market economy is the best allocator of goods and resources, yet the market does not currently extend adequately to the use of natural resources and environmental externalities:

> Environmental goods like clean air and clean water are not exchanged in
> unregulated markets: they have no price, so there is no incentive for econ-
> omy in their use. While 'free' to the individual, they are scarce to the
> community, and therefore become degraded. Pollution can be seen in this
> context as an externality – a cost of economic activity which can be imposed
> by some parties on others without compensation. Intervention to create a
> 'market' in clean air or clean water – by creating property rights in the envi-
> ronment or imposing effluent charges – will in theory internalize these
> costs and correct the market failure which leads to pollution.[93]

In order to tackle the impact of 'market failure' the environmental costs
of polluting activity are internalized through the use of market-based policy instruments enabling the costs of environmental damage to be
imposed on polluters. A key component in many market-based instruments is the use of the 'polluter pays principle' whereby costs are borne
by individual polluters as a disincentive to continue with the polluting
activity. The Paris based Organization for Economic Cooperation and
Development (OECD) has recently classified the range of economic instruments available for environmental policy into five categories:

1.   charges
2.   subsidies
3.   deposit-refund systems
4.   market creation
5.   financial enforcement incentives.[94]

In the case of waste management, the environmental economist R Kerry
Turner argues that low levels of recycling in developed economies can be
attributed to 'market failure', whereby 'the causes of the throwaway society lie in the distorted market incentives that affect both consumer and
producer behaviour'. He advocates the deployment of a range of potential
measures such as taxes on virgin raw materials and waste disposal by
landfill.[95]

The virgin raw materials market

There is much agreement between environmentalists over the need for a
tax to be imposed on virgin raw materials in conjunction with the removal
of state subsidies in order to improve the relative strength of the secondary materials market:

> Increasing the cost of raw materials is an essential step toward improving
> the efficiency of materials use and reducing waste. Virgin raw materials
> are now artificially cheap, in relation both to secondary materials and to
> other factors of production. *Prices that accounted for the real costs of using
> materials would be the single most effective incentive for source reduction, re-use
> and recycling.*[96]

Yet there has been little serious attempt to tackle this source of market failure, principally due to the political strength of the mining and raw materials processing sectors of the economy. A further cause for concern is the reliance of many Third World economies on a small number of raw materials for almost all their export earnings: Zambia, for example, receives over 95 per cent of its export earnings from copper. This raises issues concerning the operation of multinational companies and the repatriation of profits through transfer pricing and other mechanisms. Indeed, when one begins to examine ways of limiting the use of virgin raw materials it is difficult to separate the 'market failure' issue from the wider problems of how government intervention against powerful vested interests can be used for environmental protection at an international level. The 1990s are seeing a resurgence in the power of the international mining industry in order to maximise profits from the extraction of raw materials in the Third World. A number of factors are contributing to this 'new world order': the programmes of structural adjustment (including the privatization of state mining industries) and trade liberalization opening up new opportunities for multinational companies; the mounting costs of meeting environmental standards in the North; and the spread of free trade zones obviating the need for payment of royalties and other duties.[97]

Internalizing the costs of waste disposal

The contemporary market-based approach to raise levels of recycling is focused on the need to internalize the full range of environmental externalities associated with the costs of alternative means of waste disposal. The price of landfill in particular, has been targeted through a variety of mechanisms such as levies, recycling credits and proposals to extend civil liability of site operators for the future environmental impact of waste disposal. For example, a surcharge can be applied on waste disposal, and the UK government is planning to introduce a levy on all waste sent to landfill.[98] It is widely assumed that the current trend towards increasing waste disposal costs operates as an economic incentive to recycle in the sense that low landfill costs have previously acted as a disincentive to expand recycling:

> As landfill costs continue to rise because of space constraints and stricter environmental regulations, and as the high capital costs of incinerators and their pollution control technologies sap city budgets, the appeal of recycling will inevitably grow.[99]

Yet there are a number of difficulties with simplistic notions that low levels of recycling are attributable to 'market failure' in the costs of waste disposal. A first issue is that the advocacy of MBIs (market-based policy instruments) has become enmeshed in a wider political agenda for the privatization of utilities and a reduced role for local government in the provision of services, with important implications for the choice between energy and materials recovery and also between the recycling of post-consumer waste and waste reduction at source. A second complexity is that some economic policy instruments such as the UK Non Fossil Fuel Obligation (NFFO) designed to encourage the generation of electricity from non fossil fuel sources are not addressing any notional form of 'market failure'. Finally, the reasons behind the post-war expansion of the waste stream and low levels of recycling are diverse and stem in large part from the development of the global market economy and new patterns of mass production and consumption, the changing of which would imply drastic and fundamental changes in public policy.

Alternatives to the market-based approach

There are a number of concerns with the growing reliance on the use of market-based policy instruments in preference to regulatory action by government. A key issue is how the 'price signals' to influence the behaviour of producers and consumers are to be arrived at using cost-benefit analysis techniques for the monetary valuation of environmental problems. In the US at present there is a bitter policy debate over whether techniques such as 'contingent valuation' which measure people's willingness to pay for environmental protection have any useful role other than promoting certain quantitative approaches within economics. The academic statistician Zvi Griliches of Harvard University notes in the *New York Times* that 'economists just don't want to admit there is stuff that cannot be measured'.[100] If the analogy of 'money votes' is used as the basis of decision making, it is clear that many people are relatively, if not completely, disenfranchised from environmental decision making.[101] The assignment of monetary values to components of the environment also assumes a disjointed view of the environment as a collage of different pieces to which property rights may be attached and fails to determine who has rights to what and the basis on which property rights are to be allocated.[102]

A number of alternative conceptions to that of market-based approaches to environmental policy move the focus towards issues of equity and democratic public participation. In the case of waste management, the alternative non market-based approaches seek to maximize levels of recycling and waste reduction with an emphasis on the use of regulatory rather than market-based policy instruments, as set out in Table 2.4. The emergence of the 'throwaway' society is seen as the outcome of rapid and uncontrolled post-war economic growth and not as the result of 'market failure' since waste and ever higher levels of consumption are integral to a capitalist economy. Where economic policy instruments are promoted, such as differential indirect taxation on recycled and non-recycled products, this is not based on the underlying premise that environmental degradation is simply the outcome of different forms of 'market failure', but rather that the complexity and seriousness of the environmental crisis demands the employment of the full range of available policy instruments.

Table 2.4 Market-based and regulatory approaches to recycling policy

Market-based approaches	Regulatory approaches
• Differential indirect taxation for recycled and non-recycled products • The use of direct charges to households for the collection and disposal of their waste on the basis of weight or dustbin size • Marketable permit systems • Deposit refund systems • Implementation of the 'polluter pays principle' through forcing the onus for recycling back onto the producers of waste eg DSD system in Germany • Extension of civil liability for the environmental impact of waste disposal eg new EC proposals concerning the environmental and health impacts of former landfill sites • The internalization of the costs of waste for disposal through various measures such as a landfill levy or the tightening of environmental standards	• The setting of recycling targets eg New York State target of 50 per cent by the year 1997 • Legislative controls on superfluous and unrecyclable packaging • The simplification of products to facilitate recycling • The elimination of built-in obsolescence • The use of standardized returnable containers facilitated by an emphasis on regional economic systems to minimize transport distances • The promotion of strategic planning in waste management with a pivotal role for elected local government • The use of legislation to discourage the use of materials which cannot be satisfactorily recycled except as 'linearrecycling' into inferior products or their calorific value eg laminated cartons and most plastics polymers
Underlying rationale	*Underlying rationale*
• Correction of various sources of 'market failure' which are argued to lie behind low levels of recycling • Opposition to state intervention unless to facilitate operation of market-based policy instruments (MBIs) • Extension of definition of recycling to include energy recovery from waste • Emphasis on voluntary rather than include recycling programmes	• Low levels of recycling seen as outcome of the post-war 'throw-away' society • Community participation seen as integral to 'soft' small-scale technologies for waste management • Emphasis on comprehensive kerb-side collection schemes and the composting of putrescible wastes • Integration of recycling policy with wider policy aims such as employ-ment creation and the promotion of alternative forms of social and economic organization • Emphasis on waste reduction at source within the hierarchy of recycling options

Source: Adapted from Gandy, M (1993), *Recycling and Waste: an exploration of contemporary environmental policy*, Avebury, Aldershot.

3

London

Greater London covers an area of some 1600 square kilometres with a population of just under 6.8 million having fallen from a peak of over 8.6 million in 1939. Since the abolition of the Greater London Council in 1986 the city has been without any elected city-wide government with most services now provided by 33 separate municipal authorities comprising the Corporation of London in the city centre, the 32 London Boroughs, and a variety of joint boards, quangos and central government agencies. In this respect the administrative arrangements for London are unique among the world's largest cities and quite unlike both New York and Hamburg where there are powerful city-wide tiers of municipal government.

In the early 1990s some 3,700,000 tonnes of waste are collected annually by the London Boroughs and over 80 per cent of this waste is disposed of at landfill sites outside London.[1] In contrast to both New York and Hamburg, there is no imminent perception of a waste management crisis facing London and the drift towards a key role for waste incineration rather than a major expansion of recycling is occurring in the virtual absence of any public debate. Despite this quiet expansion of incineration, the promotion of recycling in order to contribute to environmental protection has assumed a important place in environmental policy for the London Boroughs and was marked nationally by the UK Government's public commitment in 1990 to raise levels of household waste recycling to 25 per cent by the year 2000 with an emphasis on the use of market-based policy instruments to raise levels of recycling.[2]

THE HISTORICAL DEVELOPMENT OF WASTE MANAGEMENT FOR LONDON

The rapid growth of London in the eighteenth and nineteenth centuries saw a marked deterioration in the quality of urban life as thoroughfares and watercourses were quickly choked with refuse and sewerage, and the medieval practice of throwing waste into the streets became intolerable. From the mid-eighteenth century onwards there were growing demands for London-wide public management of essential services such as street cleansing and water supply. In 1752, for example, Corbyn Morris noted the alarming levels of mortality in London and observed that:

> As the preservation of the health of the people is of great importance, it is proposed that the cleaning of this city should be put under one uniform public management, and all the filth be carried into lighters, and conveyed by the Thames to proper distances in the country.[3]

Yet it was not until the scientific advances of the mid-nineteenth century and the rise of the public health movement that there was widespread recognition of the connection between dirt and disease. Some indication of the squalid urban conditions in London during the mid-nineteenth century is given by contemporary accounts. For example, a survey in 1849 by the surveyor Thomas Lovick into improved 'mechanical and economical methods of street cleansing' in Soho, central London described how:

> Many of the streets and courts in which these experiments were performed, were (before cleansing) in a most filthy and insalubrious condition; the surface coated with mud, and strewn with offal and refuse of the most disgusting nature; in the interstices between the paving, and in hollows formed by its partial settlement, stagnant foetid liquids had collected, charging the atmosphere with their offensive exhalations.[4]

The late nineteenth century saw growing demands for a London-wide government to replace the increasingly chaotic system of paving commissioners, turnpike boards, vestries and other bodies administered by some 10,000 commissioners.[5] In 1847 the Health of London Association argued that it was futile to treat drainage, sewerage, street-cleansing and water supply as though they were separate responsibilities and that there had to be a comprehensive approach undertaken by an overriding central authority.[6] Initially it was the problem of coordinating highways maintenance and the direct emptying of sewage into the Thames from London's three million inhabitants which led to the formation of the Metropolitan Board of Works in 1856, representing the first functional centralization of service provision for the rapidly expanding city. In 1889 the Metropolitan Board of Works was replaced by the London County Council with jurisdiction over the new County of London composed of 29 Metropolitan Boroughs, as shown in Figure 3.1.

The period before the creation of the London County Council in 1889 was marked by a focus on street cleansing to allow free passage along highways with a very limited role for local government in the collection and disposal of waste from households. During the period of the London County Council from 1889 to 1965 each municipal authority in the Greater London area assumed the task of both refuse collection and disposal. The London County Council passed a bye-law in 1893 making it compulsory for household waste to be removed from all premises in London at least once a week. This collection of increasing quantities of waste led in turn to the need to devise satisfactory arrangements for waste disposal. During the late nineteenth century, the use of incineration plants or 'dust destruc-

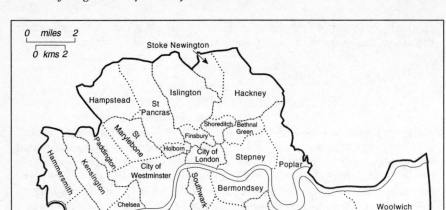

Figure 3.1 The London County Council and Metropolitan Boroughs, 1889

Source: Hoggart, K and Green, D R eds (1991), *London: A New Metropolitan Geography*, Edward Arnold, London

tors' as they were then known began to emerge at the main means of waste disposal in London, as illustrated in Figure 3.2. A significant reason for the widespread use of incineration was the Metropolitan Boroughs' important service delivery role, including the supply of electricity and in 1896 the first combined incineration and electricity production scheme in the UK began operation in Shoreditch in east London.[7] The heyday for incineration was reached in the first decade of this century, after which waste disposal was increasingly carried out by the cheaper tipping of waste at sites on the outskirts of London.

Whereas in the nineteenth century the main salvage activities had been for horse manure and other valuable items derived from street cleansing, the extension of waste collection to individual households had two main effects: firstly, the size of the waste stream for disposal began to grow rapidly as increasing numbers of households were incorporated into the collection rounds; and second, the nature of the waste stream began to change, as shown in Figure 3.3, bringing in increasing quantities of potentially useful items such as glass bottles and metal cans. The rationale for recycling now extended beyond the economic value of the materials to the perceived need to reduce the waste stream and to improve the operational efficiency of incineration plants, which could not dispose of the increasing quantities of incombustible glass and metals. There are examples of 'contractors paying a weekly sum for the privilege of sorting over the refuse and taking out materials which can be sold'.[8] Elsewhere materials were often removed by hand picking from a moving belt at waste transfer stations. A detailed description of a model plant in the early 1920s shows that a long travelling belt was used from which the following materials were removed: tin cans, waste paper, refuse of meat and fish markets (used to make manure for gardens), grease from waste meat (used for candle making and lubricating oils), and butcher's refuse and waste food (used for pig swill and poultry feed).[9]

From the 1920s onwards there were mounting difficulties for London's waste management as a result of the rapid growth of the municipal waste stream and the increasing reliance by private contractors on the indiscriminate tipping of waste. The deteriorating situation led to a government inquiry by the Ministry of Health led by the Inspector of Public Cleansing, Mr J C Dawes, which found that not only was there no

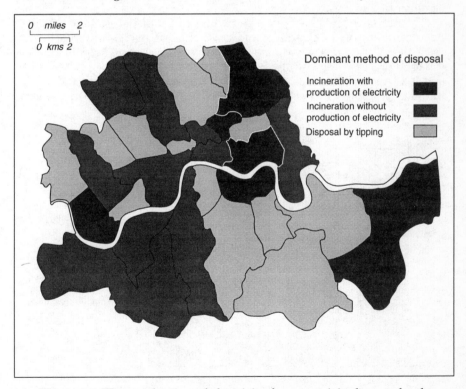

Figure 3.2 The production of electricity from municipal waste by the Metopolitan Boroughs in 1904

Source: Goodrich, W (1904), *Refuse disposal and power production*, Archibald Constable, London

effective cooperation between the Metropolitan Boroughs in order to ensure economic efficiency, but that the sites used for the disposal of waste were a serious threat to public health. The main recommendation of the inquiry was for a unified tier of local government able to coordinate waste disposal across London but no legislation was introduced by central Government.[10] By the 1950s the increasing quantities of waste produced in the London area were perceived as a matter of urgency because of the exhaustion of landfill sites and growing recognition of the environmental consequences of poorly monitored waste disposal practices. Yet it was not until the publication of the *Royal Commission on Local Government in Greater London* in 1960 that the need for improved waste management was explicitly linked to the need for wider reforms of London government.[11] Finally, under the London Government Act of 1963 waste disposal was to become the responsibility of unified London-wide government for the first time with the creation of the new GLC Public Health Engineering Department in 1965, the largest Waste Disposal Authority in Europe, while the collection of waste and street cleansing passed to the 32 new London Boroughs, as illustrated in Figure 3.4.[12]

The newly-formed GLC inherited a range of difficulties: there was little information on the quantities and types of waste (weighbridge facilities were absent from many waste transfer stations); there was no policy coordination or planning; and the waste management infrastructure was old and inadequately maintained because of impending re-organization.[13] A number of problems were identified with the 17 incineration plants inher-

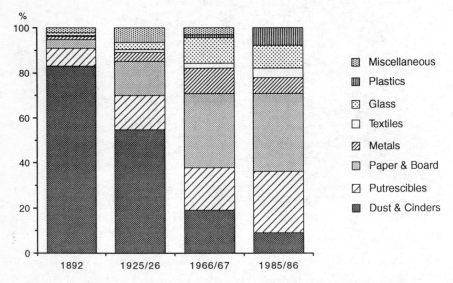

Figure 3.3 The changing composition of household waste in London

Source: Gandy, M (1993) *Waste and Recycling: an exploration of contemporary environmental policy*, Avebury, Aldershot.

Figure 3.4 The new London Borough and GLC boundaries in 1965

ited by the GLC: there was inadequate storage capacity for waste on the cramped sites; many relied on unpleasant and very expensive manual tasks such as belt picking, manual stoking and manual clinkering (removal of ash); and there was atmospheric pollution by dust, flue gases and flue grit. In addition many plants had down-rated capacities, sometimes only half that of the design capacity, owing to changes in the composition of refuse towards a lower density and an increased calorific value from paper, plastics and other combustible materials, as shown in Figure 3.3. The existing practices of tipping at landfill sites were equally found to be unsatisfactory: the ever-increasing content of paper and plastic made it impossible to avoid extensive wind-borne littering of the tip and its environment; the lower cinder and ash content (especially since the 1956 Clean Air Act) had rendered the waste more pungent and lighter in colour (less putrescibles and paper were now burnt in the home with the rise of gas and electric heating systems); the greater compaction from new mechanical as opposed to manual handling methods on the tips had resulted in a failure of waste to reach a high enough temperature to decompose sufficiently to destroy hazardous pathogens; and finally, the practice of waste disposal by landfill was increasingly unpopular with the public.[14]

Despite the initial optimism, waste management under the GLC in the 1960s proved more difficult than anticipated. The access to nearby landfill capacity for London was rapidly exhausted because of the growth of London's waste stream since 1965 and because of the impact of new legislative controls on disposal sites in response to the increasing political salience of environmental concern.[15] As a result of these developments, the waste had to be transported over progressively greater distances. Furthermore, only one of five planned new incinerators was ever built, owing to the serious technical and financial difficulties experienced with the first plant commissioned at Edmonton in north London in 1971. There were also problems with utilizing river transfer for a large proportion of inner London's waste in order to reduce the reliance on road haulage since much of London's waste was generated some miles from the river and there were limitations imposed by the location and capacity of the existing waste transfer stations along the banks of the Thames.[16]

Following the problems in expanding incineration and the use of river transfer stations, a new strategy emerged during the 1970s. This was based around the increased use of landfill at more remote sites using bulk transfer by rail rather than road to transport the waste (in contrast with the political restrictions in the use of long-distance landfill in the cases of Hamburg and New York). There were major new investment programmes embarked on, resulting in the construction of two new rail transfer stations in west and north London, each with a daily 800 tonne capacity.[17] As a result of these developments, the GLC was the only UK Waste Disposal Authority to make substantial use of rail transfer of waste with its environmental advantages over road haulage. Both new rail transfer stations were superior on both economic and public health grounds to the three rail transfer stations inherited by the GLC in 1965, which had used open top wagons with both loading and discharging in the open.[18]

A number of national legislative developments during the period of the GLC affected waste management in London. The first relates to the problem of disposing of bulky household waste such as electrical goods and garden waste. The problems arising from the fly-tipping (illegal tipping) of these kinds of waste led to the 1967 Civic Amenities Act.[19] This legislation gave local government the responsibility for providing civic amenity facilities where the public could bring items of bulky waste free of charge and by the early 1980s, some 14 per cent of the municipal waste stream in London was derived from 38 civic amenity sites, their use by the public having grown by over 12 per cent annually since their creation in 1967.[20] These civic amenity sites were later to become an integral focus of recycling activity as multi-material recycling centres with the advent of on-street collection facilities for recyclable materials during the 1980s.

Further legislative developments concerned the inadequacies of waste regulation and the disposal of hazardous and toxic wastes. In 1970, the Key Committee was set up by central government to report on the disposal of toxic waste, and recommended that the GLC should be the toxic waste disposal authority for London.[21] The 1974 Control of Pollution Act saw the introduction of three main changes for UK waste management:

1. The formal recognition of strategic policy making in waste disposal.
2. The duty of the GLC and other Waste Disposal Authorities to produce waste disposal plans.
3. The introduction of a licensing system for waste disposal sites.

By the mid-1980s, the GLC had completed a substantial investment programme in major new transfer facilities, enabling the long term coordination of waste management, improvements in economic efficiency, higher environmental standards, and the utilization of state of the art technologies in public health. The role of the GLC had evolved from simply being the Waste Disposal Authority for the annual production of 3 million tonnes of municipal waste in London, into a wider regulatory role for 17 million tonnes of industrial, construction and other wastes in the London area.

RADICAL WASTE MANAGEMENT UNDER THE GLC IN THE 1980s

Following the 1981 GLC elections a new Labour administration under the leadership of Ken Livingstone heralded a radical shift in waste management policy for London. An Environmental Panel was formed by elected councillors, and this quickly sought a strategic policy-making role in both recycling and waste management. An initial policy initiative was the emphasis on the employment creation potential of recycling. In November 1984 the *Recycling for Employment* conference was held at County Hall and it was reported that hundreds of jobs in London were provided by recycling activities. These included employees in recycling workshops, the voluntary sector, and the scrap and secondary materials industries.[22] The GLC promoted six recycling workshops by the provision of financial assistance and the granting of access to the waste stream at civic amenity sites for six inner London recycling workshops established across London in 1981 (one of which was specifically set up for the employment of people with disabilities).[23]

With the submission of the GLC's Waste Management Plan *No Time to Waste* in November 1983, some indication was given of how a new commitment to raising the level of recycling was to be achieved. The new strategy would be based on the expansion of the 'bring' system of on-street collection facilities, along with the recovery of energy from landfill gas. The GLC recognized that the provision of on-street collection facilities and recycling centres could not allow the recycling of the major fraction of municipal waste derived from putrescible kitchen and garden waste. This was seen as best achieved either through the encouragement of households with gardens to carry out composting or as an energy source from landfill gas.[24] The 1983 GLC Waste Management Plan emphasized the need to tackle the production of waste at source, with a nationally orientated waste management strategy. This marked a phase of lobbying by the GLC of both Whitehall and the European Commission. Examples include the evidence presented to the Royal Commission on

Environmental Pollution in 1984 and the European Commission in 1985, where the GLC called for the reduction and control of packaging and the increased use of returnable and reusable beverage containers based around a national beverage containers deposit scheme to be introduced by the Government.[25]

However the 1980s saw a divergence of views in the UK over the future direction of waste management policy. In 1979, the new Conservative administration began to change the emphasis of waste management policy away from the public sector: there was a reduction in funding for pollution control agencies; the nationally based Waste Management Advisory Council set up in 1974 was abolished; the enactment of Section Two of the 1974 Control of Pollution Act was postponed as a cost-cutting measure; and increasingly stringent controls on local government expenditure limited the potential for new locally funded waste management initiatives.[26] Above all, the 1980s heralded a progressively more market orientated approach to waste management in the UK with successive waves of legislation aimed at increasing the role of the private sector in municipal waste management. In contrast to national developments, senior local government officers argued that London needed a more powerful and better funded GLC Public Health Engineering Department in order to unify the collection and disposal of waste at a city-wide level as in Hamburg and New York.[27] The GLC recognized that an integrated system of waste collection and disposal, as existed in Wales and Scotland, would make materials recovery easier to coordinate[28] but the suggested modifications to the waste management framework established for London in 1965 were to be overtaken by wider political events.

THE ABOLITION OF THE GLC PUBLIC HEALTH ENGINEERING DEPARTMENT

In June 1983 the Thatcher-led Conservative government was re-elected with an increased Commons majority to its second term in office. A late and little-discussed commitment in the Conservative Manifesto had been the proposed abolition of the Greater London Council and the six Metropolitan County Councils. In October 1983, the Government produced its White Paper, *Streamlining the Cities*, stating that the GLC Public Health Engineering Department was to be abolished with a redistribution of as many of its functions as possible to the private sector:

> The responsibilities of the GLC and the MCCs for waste regulation and disposal will be transferred to the borough and district councils. The Government will wish to see that, in the setting up of the new arrangements for disposal, the maximum encouragement is given to increasing private sector participation.[29]

Opposition to the Government's proposals was expressed by almost every organization involved in the management of London's waste in both the private and public sectors, including the Confederation of British

Industry, the Civic Trust, the Hazardous Wastes Inspectorate, the Institution of Civil Engineers, the Institution of Public Health Engineers, the Institute of Wastes Management, the Federation of British Aggregate Construction Materials Industries, the National Association of Waste Disposal Contractors, the Royal Town Planning Institute, the Sand and Gravel Association Limited, the Town and Country Planning Association, the regional planning forum SERPLAN, and the Waste Disposal Engineers Association.[30] The response of the GLC Conservative Group to the White Paper remarked that:

> The GLC's record in the area of waste disposal has been a credit to local government, and its considerable advances over the years since the system was unified and streamlined have been due to economies of scale, to considerable capital investment, to a fruitful partnership with the private sector and to the expertise of its skilled and innovative professional staff... . Nowhere in the Consultation Paper is there the slightest criticism of waste disposal arrangements in London... . Devolution of waste disposal responsibilities is simply a consequence of abolition and not one of the reasons for it.[31]

The Government's aim of having a satisfactory scheme for each part of London, agreed by the 33 different municipal authorities before December 1984, was not realized. When the provisions of the White Paper were eventually enacted in the 1985 Local Government Act, the Secretary of State, Patrick Jenkin, had to use his reserve powers to set up four statutory waste disposal authorities for the twenty one boroughs which had not yet agreed on arrangements for their waste disposal. Further difficulties lay in the area of toxic waste management and the fulfilment of waste regulation responsibilities under the 1974 Control of Pollution Act. After sustained lobbying of the 1985 Local Government Bill, the regulatory functions of the GLC created under the 1974 Control of Pollution Act and responsibilities for toxic waste disposal remained intact at a London-wide level, and were passed to the London Waste Regulation Authority as a surviving fragment of the former GLC Public Health Engineering Department.

In the absence of the GLC, a number of important aspects of recycling policy have disappeared for a mix of political and organizational reasons: the promotion of recycling employment as part of a wider industrial strategy for London; the national and international lobbying role over the 1984 DTI *Wealth of Waste* report and the 1985 EC Directive on packaging; and the provision of financial assistance to charities and others involved in recycling activities. Perhaps the greatest overall impact of GLC abolition is in terms of direct democratic control over the available options for long-term waste management. In the absence of a regional waste disposal authority with a sufficiently large budget, London's municipal authorities can only afford to embark on comprehensive kerbside collection schemes by relying on either central government (which can be ruled out in present circumstances) or the private sector. The case for the re-establishment of a strategic authority for waste management in London rests with concerns such as the on-going need for a combination of strategic transport and land use planning;[32] and the minimization of waste flows through the coordination of waste transfer facilities to reduce transport costs.[33] There are also a number of specific advantages for recycling:

- the circulation of relevant information to all boroughs and the avoidance of unnecessary duplication and effort on the part of local government officers in policy analysis and production of publicity material;[34]
- the effective development of London-wide collections for materials such as plastics, for which it is difficult for individual boroughs to achieve economies of scale;[35] and
- the introduction of loss-making collection schemes on purely environmental grounds for CFCs, putrescibles and other materials, not adequately handled in the private or voluntary sectors.[36]

The current pattern of waste management in London is marked by an overwhelming dominance of landfill for waste disposal, with around 83 per cent of municipal wastes being transported by road, rail or river to large, closely monitored sites, chiefly in Essex, Kent and Hertfordshire. A further 14 per cent is burned in the former GLC Edmonton incineration plant now operated by the North London Waste Authority serving seven London Boroughs. In the year 1989–90 the level of recycling in London stood at just 2.1 per cent and Figure 3.5 shows that the level of recycling varied from nil in the Corporation of London (where there was no recycling activity) to 8.8 per cent in the largely prosperous and residential outer London Borough of Havering. Apart from Havering, only four other boroughs had a recycling rate in excess of 4 per cent, and all these boroughs were located in outer London. As for the five boroughs whose recycling rate was 0.5 per cent or less, these were all located in inner London.[37]

A major development during the 1980s, in addition to the use of on-street collection facilities, was the transformation of the role of civic amenity sites into multi-material recycling centres. In 1982, the first GLC recycling centre was opened at the Victoria Road civic amenity site in west London, handling glass, paper, cans, scrap metals, oil and textiles brought in by the public. By 1984, a further 10 of the 40 GLC-run civic amenity sites had been converted into recycling centres, as part of a major investment programme. The distribution of civic amenity facilities provides a clear indication of the extent of waste management infrastructure in individual boroughs, and in particular the scope for expanding civic amenity sites into multi-material recycling centres. Inner London Boroughs such as Hammersmith and Fulham, for example, do not have any civic amenity sites, which in many other boroughs have been successfully turned into multi-material recycling centres, recovering scrap metals, oil, textiles, CFCs from refrigerators and many other items. The importance of civic amenity sites used as recycling centres is also suggested by Figure 3.6. This shows how the range of materials recovered from the waste stream varied from under four materials in some inner London Boroughs, to over

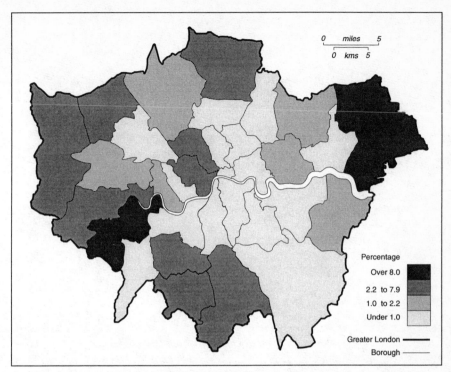

Percentage

Over 8.0

2.2 to 7.9

1.0 to 2.2

Under 1.0

Greater London ———

Borough ———

Figure 3.5 The level of recycling in the London Boroughs, 1989–90

Source: Gandy, M (1993), *Waste and Recycling: an exploration of contemporary
environmental policy*, Avebury, Aldershot.

ten different kinds of materials in the better equipped boroughs, with both
recycling centres and a wider range of on-street collection facilities available.

The wide variations in the level of recycling across London can be
accounted for by the extent of recycling facilities, determined primarily by
a combination of urban congestion, land values and the distribution of
recycling infrastructure inherited by the London Boroughs from the GLC
in 1986. The congested inner London Boroughs face a physical barrier to
recycling in terms of the lack of space along highways and in common
areas for the location of collection facilities. Physical congestion both in
smaller homes and flats, combined with cramped common areas and
pavements, makes the introduction of more on-street collection facilities
extremely difficult.[38] The main means by which materials are separated
from the waste stream in London is by the 'bring' system of recycling,
comprising on-street collection facilities (mainly for glass, paper and metal
cans) and civic amenity type facilities used as recycling centres for a wider
range of recyclable materials. There are only a few exceptions to the use of
the 'bring' system in London: a loss-making kerbside collection scheme

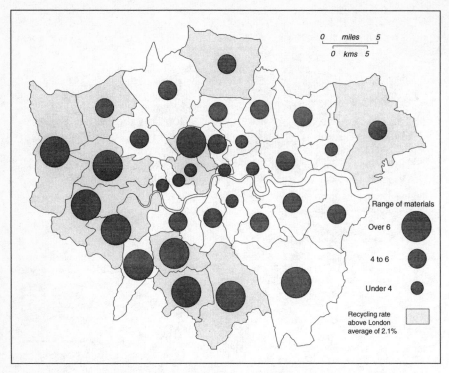

Figure 3.6 The range of materials recycled in London, 1989–90

Source: Gandy, M (1993), *Waste and Recycling: an exploration of contemporary environmental policy*, Avebury, Aldershot.

for paper in the London Borough of Havering; voluntary sector kerbside collections of clothes and other unwanted household goods in most boroughs; and the recovery of ferrous metals by magnetic separation at the Edmonton incineration plant in north London.

There are now no major schemes for the recovery of putrescibles from household waste in London. The question of composting was considered by the GLC in 1985, but centralized production was ruled out for two main reasons: there was doubt over its cost-effectiveness; and there was no satisfactory recent experience of running centralized composting programmes. As a result of these uncertainties it was concluded that the best option would be the promotion of home composting activities.[39] The decline in the practice of composting waste originates from the increasing presence of heavy metals derived from the growing proportion of toxins in the municipal waste stream. This was the reason behind the closure of the Sunbury-on-Thames composting plant inherited by the GLC in 1965. A number of boroughs have sought to encourage households with gardens to carry out their own composting activities but there is no direct involvement of local government in the form of separate organic waste collection services or the provision of centralized composting facilities. In

1991, the North London Waste Authority (serving seven boroughs) began a scheme for the composting of waste from parks, but this has not been extended to collections from individual households.

THE POST-WAR COLLAPSE OF PAPER RECYCLING

When the local government system for London was reorganized in 1965, 25 of the new London Boroughs inherited municipal collection services for waste paper, as shown in Figure 3.7. When the GLC took over the responsibility for waste disposal they introduced a rebate for the London Boroughs to encourage the survival of existing kerbside collection services based on the marginal cost savings for waste transport and disposal, calculated on a separate basis for the individual boroughs. Consequently, the boroughs in inner London which used waste transfer stations and incurred high costs of transport in addition to their tipping fees, received higher levels of rebate than the boroughs in outer London which delivered their waste directly to landfill sites with their own collection vehicles.[40] The GLC also carried out price negotiations on behalf of the London Boroughs with Thames Board Mills and Merton Paper Mills. Yet

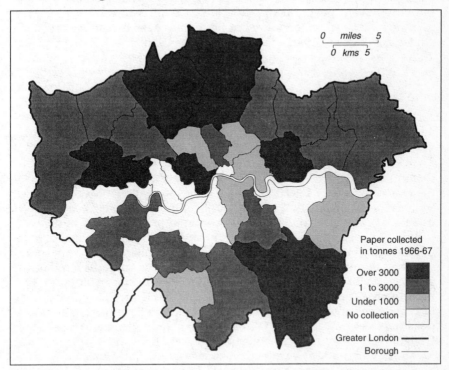

Figure 3.7 Waste paper collections by the London Boroughs in 1966–67

Source: Greater London Council

in spite of the GLC rebate and negotiated price contracts, the quantity of paper collected by the London Boroughs declined steadily and by 1975 the number of boroughs running kerbside collection schemes for waste paper had fallen to 16. The main reason for this decline in kerbside paper collections by local government was attributed to the increasing labour costs of collection and the weakness of the secondary materials market for paper.[41]

The collapse of many established recycling schemes in the 1970s led to intense debate over the role of central government in maintaining recycling in the national interest. As early as 1973, the Paper and Board Federation had been lobbying central government to undertake intervention buying in order to stabilize the waste paper market and there were discussions over the possibility of introducing joint government and industry schemes, as already successfully established in Japan and the Netherlands, but these never materialized.[42] The Labour MP Brynmor John called for a publicly owned recycling industry, since 'private enterprise has failed, except in scrap metal, to get a recycling industry into being on its own account' and he suggested that the best solution would be the construction of twenty regional centralized sorting plants, to be owned by and operated within the public sector.[43] In September 1974 the minority Labour government published its White Paper *War on Waste: A Policy for Reclamation*, endorsing a nationally based waste reclamation strategy, recognizing the potential difficulties facing individual municipal authorities, voluntary groups and firms handling the collected materials.[44] By December 1974, however, the economic slowdown heralded a rapid collapse in the waste paper market, and the industry called for £10 million of state aid under Section Eight of the 1972 Industry Act.[45] By early 1975 the GLC was calling for urgent government assistance to prevent the collapse of paper collection services in the London Boroughs[46] and the GLC Public Services Committee reported in 10 June 1975 that Thames Board Mills were refusing to accept any more paper because of their huge stockpile.[47] Just two days later, for example, Hounslow council made an emergency decision to stop its waste paper collection service with immediate effect.[48] Yet at this critical moment in 1975, as a consequence of mounting political and economic problems and ministerial reshuffles, recycling quickly slipped down the public policy agenda. The deteriorating situation had led to the disposal of collected paper at landfill sites and there were urgent investigations into ways of reducing the cost of waste paper collections. The mid-1970s saw calls for voluntary sector involvement in collection to be quickly expanded in order to cut the rising labour costs[49] and the termination of municipal run services created furious local rows over the loss of 'trailer money' for drivers with waste paper trailers behind their waste collection vehicles.[50] In addition to calls for greater voluntary sector involvement, there were proposals to examine alternative and cheaper means of collecting paper by encouraging the public to bring paper to civic amenity sites.[51] This harnessing of the potential for public participation was later to form the basis for on-street collection facilities such as bottle banks and multi-material recycling centres.

The more recent fluctuations in the price of paper since the mid-1980s have again had a disastrous impact on the recycling of paper from household waste. This can be attributed to several factors:

- Household waste paper consists of predominantly lower grades, such as newspapers, yet demand is strongest for higher grades of office and computer paper.
- Household waste paper is frequently a mixture of different kinds of paper and card, which are difficult to sort and command a lower price than sorted grades.
- The paper is often contaminated by laminates and other materials.
- The period since 1988 has seen a rapid increase in the collection of waste paper by households, swamping an already fragile market for mixed and low grade paper.

A key problem has been an accumulated glut of over 800,000 tonnes of waste paper in the UK, the flooding of the market with cheap imports from North America and Germany following the implementation of mandatory recycling legislation, and the continuing uncertainty surrounding plans to build two new paper mills at Aylesford in Kent and Gartcosh in Scotland because of unfavourable macro-economic conditions for investment.[52]

The period since 1988 has seen the disappearance of many schemes run by municipal authorities, the private sector, and the voluntary sector and the impact of falling paper prices has affected both the cheaper 'bring' system of on-street collection facilities for paper as well as kerbside collection schemes. By the mid-1980s the only municipal run kerbside collection scheme for paper still operating was based in the borough of Havering in outer London, and this had become a financial disaster since 1988, dwarfing all other recycling income in the borough from other materials, including glass and scrap metals.[53] Despite the chaotic impact of the secondary materials market on paper recycling, the UK Government have reiterated their opposition to any state intervention to assist recycling. In January 1990, for example, the government minister with responsibility for recycling, David Heathcoat-Amory, stated that there were no plans to help charities and other organizations thrown into financial difficulties by the fall in waste paper prices.[54]

The recovery of low grade paper from households illustrates that factors outside the control of local authorities, such as the state of the secondary materials market and the absence of sufficient capacity in the paper recycling industry, are far more significant obstacles to paper recovery than the willingness of the public to participate. The problems with paper recycling have created confusion and disillusion on the part of the public, and exasperation on the part of local government officers inundated with public demands for paper recycling facilities. So serious has the impact of the slump in the waste paper market been, that this has now become one of the justifications behind the expansion of the burning of municipal waste to produce electricity as an alternative to materials recycling.[55]

THE RISE OF THE 'BRING' SYSTEM FOR THE RECYCLING OF GLASS

The recovery of glass from household waste declined steadily over the post-war period to reach a nadir in the mid-1970s. This was due to a combination of the high labour costs of hand picking at waste transfer stations and the absence of investment in new mechanical recovery plants for London.[56] Many of London's waste transfer stations were on cramped sites where the introduction of conveyor belts for the manual or mechanical sorting of waste would have been impractical. The proportion of packaging derived from returnable glass bottles had also declined sharply since the 1960s as other containers which were lighter, cheaper and easier to stack and transport had entered the market[57] and this decline in the use of returnable containers led to direct criticism of the glass industry by the GLC:

> The re-use of glass bottles and jars is a field in which we have gone backwards over recent years because of a deliberate policy of industry, namely, designing for scrapping.[58]

The shift from the 1970s methods of recycling to the 1980s use of 'bring' systems dependent on public participation can best be illustrated through the case of glass. The glass industry had been carrying out research into re-using glass cullet (broken glass) since 1973 but had not yet devised an economically viable scheme for the separate collection of glass.[59] In 1975 there were discussions between the GLC and United Glass over the possibility of encouraging the public to deliver used bottles to civic amenity sites since it was concluded that any system reliant on municipal authorities to collect, sort and clean bottles would be too expensive. The emergence of the widespread use of on-street collection facilities for glass began with the launch of the bottle bank scheme by the Glass Manufacturers Federation in 1977 and by 1982 15 boroughs had established a system of on-street collection facilities for glass[60] and the tonnage of glass recycled in London grew steadily, as suggested by Figure 3.8.

The use of bottle banks has not been without difficulties. In inner London, there are problems in locating a sufficient density of bottle bank facilities in congested boroughs where most households do not have a car.[61] In contrast, in outer London Boroughs such as Bromley, there is the difficulty of low population densities creating problems in terms of providing sufficient access to facilities and reaching site usage targets.[62] A further difficulty with the operation of bottle banks is the contamination of glass at the collection points. The main sources of contamination are the

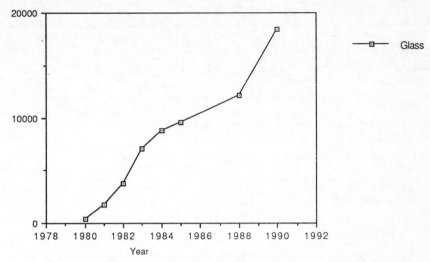

Figure 3.8 Glass collected in the London Boroughs, 1980–90

Source: Gandy, M (1993) *Recycling and Waste: an exploration of contemporary environmental policy*, Avebury, Aldershot.

mixing of different colours, especially clear with green glass, since the colour contamination thresholds are higher for clear glass in comparison with green glass. In the case of Westminster in central London, the degree of contamination with other materials had led to the extensive dumping of glass from bottle banks at landfill sites.[63] The problem in Westminster derived, in particular, from the presence of broken crockery from hotels, restaurants and other establishments and is therefore related to the high proportion of commercial waste within the borough's municipal waste stream. The adoption of either two- or three-colour separation systems for public glass collection facilities is becoming an increasingly important consideration because of growing difficulties arising from the potential saturation of the market for green and mixed glass, illustrated in Table 3.1. Green and mixed glass forms the dominant fraction of recovered glass cullet, yet the demand is strongest for clear and brown glass in the glass industry. One possible solution is the use of a mixed glass colour for all containers in order to remove this potential hindrance to further recycling, but this would require national and EC level legislation and would be strongly opposed by the glass industry.

The experience of London suggests that there is an environmental trade-off in the use of bottle banks between the global environmental goals of recycling policy and the local environmental impact of their operation. Table 3.2 shows that the operation of bottle banks has been associated with a number of problems including noise from their usage and emptying at unsocial hours, and the overflowing of broken glass onto

Table 3.1 The pattern of glass consumption and glass recycling in the UK

	Clear	Brown	Green
Glass consumption in tonnes	1,250,000	250,000	280,000
per cent	68.5	15.3	16.2
	Clear	Brown	Green*
Glass recycled from post-consumer waste in tonnes	47,300	11,300	129,300
per cent	25.1	6.0	68.8

* The green category includes mixed glass

Source: British Glass and the UK Department of the Environment (1991), *Waste Management Paper No.28: Recycling.*

the street. Further difficulties are visual intrusion (particularly from the older metal containers); the encouragement of litter and fly-tipping in the vicinity; and the periodic traffic congestion created by people using the facilities and the collection vehicles. The operation of bottle banks has also been subject to mounting criticism from the public, along with critical coverage within the London press.[64]

THE UNDERLYING BARRIERS TO RECYCLING IN LONDON

Although the largest fractions of London's municipal waste are paper and putrescibles, the main types of materials recovered are paper, glass and metals, reflecting their ease of separation and collection, along with the state of the established secondary materials markets. Table 3.3 shows that the most important barriers to recycling can be divided into three groups:

1. the weakness of the secondary materials market;
2. the cost of new recycling facilities and kerbside schemes, in the context of restrictions on local government expenditure; and
3. the need for greater controls over products and processes, particularly the 35 per cent by weight of waste, constituted by packaging.

It is apparent that the most important barriers to recycling lie effectively outside the control of local government, which helps to explain why recycling policy in the boroughs is only weakly linked to the level of recycling, as illustrated in Table 3.4. Much of the current discussion of recycling policy in London is marked by an over-emphasis on individual boroughs as if differences in recycling simply reflect the degree of commitment of administrations to environmental policy.[65] Yet this ignores the greater

Table 3.2 Problems experienced with the operation of bottle banks in London

	Important or Very important		Not important		Not known or data withheld	
	%	No of Boroughs	%	No of Boroughs	%	No of Boroughs
Complaints from local residents	75.0	24	15.6	5	9.4	3
Periodic overflowing of glass onto the street	68.8	22	15.6	5	15.6	5
Capital costs of providing facilities	62.5	20	25.0	8	12.5	4
Unreliability of cullet collection	50.0	16	37.5	12	12.5	4
Organisational and logistical problems	46.9	15	31.3	10	21.9	7
Limited public participation	34.4	11	46.9	15	18.8	6
Lower market value of green and mixed glass cullet	15.6	5	34.4	11	43.8	14
Gaining of planning permission	15.6	5	65.6	21	18.8	6

Source: Gandy, M (1993), *Recycling and Waste: an exploration of contemporary environmental policy*, Avebury, Aldershot.

prosperity and less pressing social and economic problems facing outer London Boroughs in comparison with deprived inner London Boroughs. This picture of relative disadvantage is amplified further when the distribution of recycling infrastructure, car ownership and availability of suitable locations for on-street collection facilities is taken into consideration.

Table 3.3 shows that limited public participation is perceived as the least important factor affecting recycling, with the exception of the abolition of the Community Programme employment creation scheme, which adversely affected some voluntary sector initiatives.[66] There is little evidence of any unwillingness of the public to participate in recycling across London with a favourable public response in both the wealthiest and poorest boroughs, yet there has been no attempt in London to extend a mandatory kerbside source separation scheme to deprived high rise housing estates as in New York City. The role of technical barriers to recycling is also of minimal significance, reiterating the fact that present rates of recycling do not approach the technical barriers to materials reclamation. Indeed in the case of plastics, which pose a range of potential technical difficulties for recycling, the greatest current threat to the emergence of plastics recycling in London is the flood of plastics imports from Germany as a result of the implementation of the 1991 German Packaging Ordinance.[67] It is also of interest that the low cost of waste disposal by landfill is not perceived by local government officers[68] as a significant barrier to recycling, since this is identified in much of the literature and also in official

Table 3.3 The perceived barriers to recycling in the London Boroughs

	Important or very important		Not important		Not known or data withheld	
	%	No of Boroughs	%	No of Boroughs	%	No of Boroughs
The weakness of the secondary materials market	84.4	27	3.1	1	12.5	4
The need for more on-street collection facilities	78.1	25	6.3	2	15.6	5
Restrictions on local government expenditure	71.9	23	3.1	1	25.0	8
The high labour costs of kerbside collection	71.9	23	6.3	2	21.9	7
The need for legislation to control the packaging industry	71.9	23	6.3	2	21.9	7
The high capital costs of kerbside collection	71.9	23	9.4	3	18.8	6
The need for legislation to ensure production of recyclable and durable products	71.9	23	9.4	3	18.8	6
The need for more recycling centres	68.8	22	6.3	2	25.0	8
The lack of coherence in the Government's overall strategy to increase recycling	65.6	21	6.3	2	28.1	9
The need for more recycling officers	62.5	20	15.6	5	21.9	7
The capital costs of centralized recovery	59.4	19	12.5	4	28.1	9
The need for more environmental education	59.4	19	12.5	4	28.1	9
Lack of public knowledge of available facilities	56.3	18	25.0	8	18.8	6
The need for a strategic authority for London	53.1	17	18.8	6	28.1	9
The impact of the Community Charge	46.9	15	9.4	3	43.8	14
Limited partnership with voluntary sector	46.9	15	21.9	7	31.3	10
The impact of CCT in waste collection	46.9	15	21.9	7	31.3	10
The need for staff training and expertise	46.9	15	31.3	10	21.9	7
The potential impact of the Environmental Protection Act	43.8	14	9.4	3	46.9	15
The need for legislation to ensure that householders segregate their waste	43.8	14	25.0	8	31.3	10

Table 3.3 continued

	Important or very important		Not important		Not known or data witheld	
	%	No of Boroughs	%	No of Boroughs	%	No of Boroughs
The technical barriers to recycling	43.8	14	28.1	9	28.1	9
The need for greater usage of civic amenity sites	43.8	14	31.3	10	25.0	8
The impact of high interest rates on costs and investment programmes	40.6	13	18.8	6	40.6	13
The low cost of landfill	40.6	13	31.3	10	28.1	9
The lack of support from other officers	37.5	12	28.1	9	34.4	11
The lack of support from elected members	37.5	12	34.4	11	28.1	9
Workforce resistance to new policies	34.4	11	34.4	11	31.3	10
Limited public participation	34.4	11	43.8	14	21.9	7
The abolition of the Community Programme	15.6	5	43.8	14	40.6	13

Source: Gandy, M (1993), *Recycling and Waste: an exploration of contemporary environmental policy*, Avebury, Aldershot.

government documents as the primary source of 'market failure' which has hindered the development of recycling. This is significant because the proposed UK landfill levy has been portrayed as a 'lifeline to recyclers' but the impact is predicted to encourage the expansion of incineration rather than materials recycling.

MARKET-BASED RECYCLING POLICY

The focus of contemporary market-based recycling policy in the UK has been on the internalization of the costs of waste disposal through the UK government's mandatory introduction of recycling credits under the 1990 Environmental Protection Act replacing the existing voluntary system of rebates recognized under the 1974 Control of Pollution Act:

> Market forces are the best way to deliver a sustainable approach for waste and recycling for the long term.... Recycling credits are payments by local authorities to those who collect materials for recycling.... The credits will provide a market incentive to recycle waste.[69]

Waste disposal rebates were first introduced by the GLC for waste paper collections by London Boroughs and from 1982 for glass following the

Table 3.4 Correlation analysis of levels of recycling in London, 1989–90

	Per cent recycling rate
Proportion of households living in flats maisonettes or rooms	–0.524 **
Population density	–0.586 **
Level of socio-economic deprivation (Z-score of four indicators)	–0.585 **
Level of higher and further educational qualifications	+0.170
Population per bottle bank	–0.367 **
Proportion of waste derived from civic amenity sites	+0.526 **
Level of waste disposal rebate (recycling credit)	+0.271
Sophistication of recycling policy (Policy Index measure)	+0.286

Significance levels

0.05 *
0.01 **

All coefficients refer to log data

Source: Gandy, M (1993), *Recycling and Waste: an exploration of contemporary environmental policy*, Avebury, Aldershot.

increasing participation in the Bottle Banks Scheme launched by the Glass Manufacturers Federation in 1977. With the abolition of the GLC in 1986, three of the four statutory Waste Disposal Authorities have been granting rebates to their constituent boroughs and the calculation of waste disposal savings has also been carried out by most boroughs within voluntary waste disposal groups. With the 1990 Environmental Protection Act, rebates or 'recycling credits' now form the key element in the Government's market-based strategy to raise the level of recycling by the internalization of the costs of waste disposal, but there remain a number of doubts as to whether the use of recycling credits will bring about a substantial increase in the extent of recycling. Table 3.4 suggests that the level of waste disposal rebates appears to have no clear impact on the extent of recycling activities in London. Rebates and credits could be conceived as simply an 'end of pipe' attempt to introduce the 'polluter pays principle' into waste management. The role of credits and rebates has been over-emphasized, because the money is simply circulating between different tiers of local government and may not even enter recycling budgets as a form of incentive. In the year 1989–90, for example, the incorporation of rebates from waste disposal savings into expenditure on recycling activities was undertaken in just six boroughs, whereas in the other boroughs the rebate money simply entered the general rate fund.[70] Doubt over the effectiveness of rebates is also provided by historical evidence of the decline in local authority waste paper collections in the 1970s, despite the introduction of a rebate as a result of rising labour costs and a deteriorating secondary materials market for waste paper.[71]

Both in Hamburg and New York, the issue of rebates did not arise, because the same tier of local government is responsible for both the collection and disposal of waste, and any savings in the costs of waste management accrue directly to one overall budget for waste management. This suggests that the issue of rebates is simply an anomaly arising out of the administrative separation of waste collection and disposal. This is borne out by the fact that the legislation on recycling credits will not apply to Scotland and Wales, where the same tier of local government carries out both the collection and disposal of waste. The experience of London suggests that the use of rebates or credits, both now and in the past, has been a relatively ineffectual policy tool in comparison with the need to strengthen the secondary materials market and fund the provision and operation of recycling facilities. It is of interest that the non market-based economic instrument represented by the introduction of the Non Fossil Fuel Obligation (NFFO) subsidy for non fossil-fuel sources of electricity is already having a much more significant role in waste management than rebates have ever had. This impact is not leading to higher levels of materials recycling but towards increased energy recovery from landfill gas and incineration plants.[72]

On the basis of the perceived barriers to recycling identified in Table 3.3, a recycling rate of 25 per cent in London, in order to fulfil the UK Government's target, would require increased state intervention in the production process to control packaging and other components of the waste stream, the extension of fiscal and other measures to the secondary materials market and recycling industry, and also higher public expenditure in order to finance comprehensive kerbside collection schemes. The concern over the cost of kerbside collection schemes is also reflected nationally, with a recent survey suggesting that 84 per cent of UK local authorities have no intention of introducing kerbside collection schemes because of the costs involved,[73] yet most research suggests that recycling rates in excess of around 20 per cent cannot be achieved without the operation of some form of kerbside collection scheme because of limited rates of materials recovery with 'bring' systems.[74] A new development of great interest is a kerbside collection scheme launched in 1993 in the centrally located London Borough of Kensington and Chelsea where recyclable materials are being collected with routine waste collection services and sorted at the Cremorne Wharf waste transfer station on the River Thames. It is envisaged that this programme, carried out by the private sector under contract, will result in lower waste management costs for the borough after the inclusion of the recycling credit payments from the Western Riverside Waste Authority and the introduction of the proposed landfill levy.[75] However, it remains to be seen whether this represents a satisfactory organizational solution to the high labour costs of comprehensive recycling systems and whether there will be markets available for the collected materials to help offset the running costs of the scheme.

Despite the enactment of the 1990 Environmental Protection Act, there is still no coherent policy for recycling and waste management at a national level, which addresses the full range of barriers to higher levels of materials recycling. Indeed, Table 3.3 shows that the need for a coherent national waste management and recycling policy is identified by twenty one boroughs as necessary for increasing levels of recycling. A number of developments since the passage of the Environmental Protection Act suggest that the UK Government has begun to modify earlier resistance to any national level state involvement in recycling policy. Examples include the promotion of a voluntary 'green levy' on tyre producers; the introduction of a landfill levy; pressure on newspaper publishers to raise the level of recycled fibre usage to 40 per cent by the year 2000; and the setting up of a new Advisory Committee on Business and the Environment to investigate the creation of markets for recycled products. In a sense, this drift towards some form of financial assistance and limited intervention brings national recycling policy back full circle to where it was in 1974 with the publication of the *War on Waste* Green Paper and the first tentative attempts to build a national recycling and waste management strategy.[76]

THE DEMUNICIPALIZATION OF WASTE MANAGEMENT IN LONDON

The political challenge to municipal waste management in the UK can be traced within the general conflict between central and local government. This arose from a divergence of political priorities with the collapse of the post-war consensus and also through attempts to control public expenditure. In a 1984 study, the UK Audit Commission found that waste collection in England and Wales was costing in excess of £500 million a year, making it one of the most expensive local government services.[77] The report recommended that cost savings could be achieved by a variety of measures:

- The elimination of existing time standards, such as 'task and finish' practices and ineffective bonus schemes, constituting up to 30 per cent of labour costs.
- Changing vehicle and crew levels.
- Changing collection methods for faster and easier collection of dustbins from households.
- Improving vehicle utilization rates and maintenance.
- The introduction of competitive tendering for refuse collection.

In the early 1980s, competitive tendering began to emerge as something of a panacea to the New Right and there was increasing interest in extending the role of the private sector into waste collection, based on the comparative experience of refuse collection in Europe and especially the US.[78] Competitive tendering was seen as enabling a reduction in local government expenditure; the weakening of public sector unions, which had become so unpopular during the 1970s; and providing an opportunity to increase the role of the private sector in the provision of public services.[79]

Under the 1988 Local Government Act, compulsory competitive tendering was extended to a range of local government services including catering, leisure management and waste collection for the first time, releasing £30 billion of potential business to private sector companies, should they win the contracts.[80] In the period since 1988, the private sector has tendered successfully for a growing number of refuse collection and street cleansing contracts in London. In Bethnal Green, for example, the waste collection contract went to the Anglo-French Cory Onyx (which has also won contracts in the boroughs of Bromley and Camden). In 1990 Cory Onyx split into two companies: Cory Environmental, now collecting waste for Bromley and Bethnal Green, and Onyx UK with waste collection contracts for the Corporation of London, Brent and Camden. Onyx UK is a wholly owned subsidiary of Compagnie Generale d'Enterprises Automobiles, a branch of Compagnie Generale des Eaux (CGE), Frances's fifth largest company. The European companies have established experience in tendering for contracts and they have financial backing from larger parent companies taking a longer term view than their UK competitors. They have also built up a high degree of technical expertise in the field of municipal services, not only in waste management but also areas as diverse as water, energy, leisure management, homes for the elderly and mortuaries.[81]

In the 1990s the process of privatization has been extended from waste collection to waste disposal in London through the separation of local government waste disposal services into Local Authority Waste Disposal Companies (LAWDCs). The origin of the formation of LAWDCs can be traced to the early 1980s debate over the use of compulsory competitive tendering to extend the private sector role in waste management to not only the collection but also the disposal of waste. A study commissioned by the UK Department of the Environment in 1981 concluded that the costs of municipal waste management could be significantly reduced by the forming of joint venture companies to handle all or part of the waste disposal functions of local government. In particular it was suggested that the greatest cost savings might be achieved by contracting out both waste collection and waste disposal simultaneously to the same contractor.[82] Although the rationale for reducing the public sector role in waste disposal was being clearly established by central government in the early 1980s, there remained a number of uncertainties over whether the private sector had the capacity to take on a major operational role in waste management, particularly for large urban areas. It was also recognized that waste disposal could not be fully privatized because of the regulatory and licensing aspects of waste management, so that attention was focused on the operational alternatives for service delivery.[83]

In January 1989 the UK Government produced a White Paper entitled *The Role and Function of Waste Disposal Authorities*, as the result of a wide-ranging review of waste management policy in England and Wales, initiated by the then Secretary of State for the Environment, Nicholas Ridley, in September 1986.[84] The White Paper argued that local authorities should be divested of their operational role on the premise that the dual operational and regulatory role of local authorities was not in the best interests of high environmental standards. These proposals were incorporated into Part 2 of the 1990 Environmental Protection Act by placing the operational aspects of waste disposal authorities into an 'arms length' local authority waste disposal company, called a LAWDC. As a result of this legislation, the four statutory Waste Disposal Authorities in London will be abolished in their current form and the operational side of their activities will form LAWDCs, whose shareholders will be the constituent boroughs.[85] The Waste Disposal Authorities will remain as a purely client side organization, seeking tenders from the newly formed LAWDCs in competition with private sector waste disposal companies, not only for the disposal of waste but also other operational activities including the running of civic amenity sites and waste transfer stations. However the structure of the LAWDCs will make it impossible for them to raise the necessary capital in order to improve facilities and raise environmental standards and will also render them at a competitive disadvantage in relation to larger private sector waste disposal companies with their availability of greater resources.[86] Not surprisingly, the private sector waste disposal industry represented through the National Association of Waste Disposal Contractors has welcomed the formation of LAWDCs.[87]

There is also concern in London over the potential loss of local authority control of civic amenity sites and recycling centres, when these are put out to tender. This is understandable, since for many London Boroughs, the civic amenity sites form a major focus of recycling activities and also provide a source of income to local authorities from the sale of materials such as scrap metals.[88] It seems probable that the better equipped and better run civic amenity sites will be especially susceptible to outside tendering bids. This places local government in a dilemma, since if money is spent on developing good civic amenity sites with new plant and equipment but control of the site is then lost, municipal authorities will still have to pay for the disposal of materials collected there, but without any income from the sale of materials.[89] The prospects for civic amenity sites and recycling centres may be especially poor in inner London for two main reasons. Firstly, many inner London sites handle larger proportions of low value commodities such as paper, the value of which is less than their notional savings in waste disposal. These would no longer apply if a waste disposal company was running both civic amenity sites and waste disposal.[90] Secondly, it is difficult to conceive how civic amenity type services with a low rate of return could survive in the face of high land values, and private sector pressures for other land uses with a higher rate of return.[91]

In summary, it is likely that the introduction of competitive tendering in waste disposal will be far more complex and uncertain in outcome than for waste collection. Indeed, the creation of LAWDCs can be viewed as simply a phased privatization of waste management in the UK, as the LAWDCs are displaced by larger competitors in the private sector. These concerns are borne out by more recent evidence which suggests that many UK municipal authorities are not forming LAWDCs but are simply selling off their assets to the private sector amid continuing confusion over the arrangements for capital investment.[92] This is the case in the voluntary South West London Waste Disposal Group, where the boroughs of Bromley, Croydon, Kingston, Merton and Sutton have opted simply for the privatization option, and will not be setting up a LAWDC. The public sector is likely to retain its regulatory role (though in an uncertain form) but the decline of the operational role of the public sector will have important effects on a range of recycling and waste management activities currently undertaken by municipal authorities, such as the operation of civic amenity sites, recycling centres and waste transfer stations. The process of competitive tendering is altering the relationship between the different agencies involved in the provision of services, and has important implications for the manner in which policies are developed and implemented. Its significance for recycling is somewhat paradoxical, since it has introduced the possibility for change in waste management but simultaneously increased the pressure to reduce costs. Its impact on policy is arguably a narrowing of the terms of reference to criteria of economic efficiency and also a diminution in the legitimate scope of public policy, as the planning for operational aspects of waste management is handled increasingly by large private sector companies on a profit basis:

> The competitive tendering principle, which now drives local government, is based on the assumption that the public are consumers only. Their sole legitimate interest is in the quality and price of services. They are not supposed to have any stake in those services, or any control over how they are provided, who is employed, or on what terms.[93]

THE FUTURE OF LANDFILL

Some of London's Waste Disposal Authorities and larger private sector waste management companies have their own landfill sites which they will continue to use for at least another ten years. However, it is predicted that by the year 2000, nearly half of the London and South East region may be facing a shortage of landfill space for household waste and the only significant volumes of landfill space will be in the outer north-west part of the region.[94] In addition to the predicted future shortages of landfill, there are a number of operational and technical changes taking place in the use of landfill for waste disposal. For technical control reasons, in order to raise environmental standards, there is a trend towards larger landfill sites. As a consequence, these sites are being operated by fewer and larger private contractors, with a gradual diminution of the direct operational activities of the waste disposal authorities. In this respect, the formation of LAWDCs in the Environmental Protection Act will accelerate this trend already underway in waste management.[95]

One development is the systematic extraction of gas as a fuel source from landfill sites, now integral to the Government's market-led strategy to raise the proportion of electricity generated from non fossil-fuel sources to 24 per cent by the year 2025.[96] Energy production from landfill gas has been carried out since 1982 at the former GLC Aveley landfill site in Essex[97] and the trend towards landfill gas extraction from the larger landfill sites serving London is continuing: in the summer of 1990, Cory Environmental Ltd announced that it would be generating 3.7 megawatts (MW) of electricity from late 1991 at its vast landfill site at Mucking, Essex, and has signed a deal with Eastern Electricity.[98] In the UK as a whole, the extraction of landfill gas has increased steadily during the 1980s for a number of reasons: the larger landfill sites in use in the 1980s have improved the economics of bio-gas recovery: the higher cost of landfill has encouraged operators to control pollution profitably; the design and following electricity privatization, the Non Fossil Fuel Obligation (NFFO) was introduced in 1990, which instituted a levy on fossil sources, thereby improving the financial viability of renewable and non-fossil sources. There is also the potential to integrate gas recovery into schemes for the anaerobic fermentation of putrescible waste and the recycling of sewage sludge, which may expand as a result of the privatization of water and the stricter controls on dumping of sewage at sea following the 1990 Conference on the North Sea and new EC Directives.

The operation of landfill sites for London's waste is increasingly handled in large private sector companies specializing in waste management and the provision of other municipal services. The critical long-term issue is what will happen to waste management in London in the twenty first century after many of the current landfill sites are exhausted or too distant for economic use. The next section suggests that rather than an inevitable increase in efforts to recycle the waste stream, as is widely assumed both by Government and by environmentalists, the most likely development will be an increase in the profitable burning of waste with electricity generation.

THE RE-EMERGENCE OF INCINERATION

In the 1970s, the cost of incineration was consistently seen as higher than other alternatives to landfill for London, such as the production of refuse derived fuel[99] but the use of incineration as a waste disposal option in London is now re-emerging for the first time in over twenty years. This can be attributed to four main developments:

1. The rising cost and future uncertainty of landfill, especially for inner and south London.

2. The improved profitability of the former GLC Edmonton incineration plant.
3. The introduction of the NFFO subsidy for non fossil-fuel sources of energy.
4. The price collapse in the waste paper market and other difficulties with materials recycling, forcing a re-evaluation of different recycling options.[100]

The first incineration scheme to emerge for London has been put forward by the South East London Waste Disposal Group (SELWDG) now serving the boroughs of Greenwich and Lewisham. This involves the construction of a new incineration plant on derelict land in Deptford in preference to long distance landfill, and will be the first incineration plant to be built in the UK since 1976. The project is managed by South East London Combined Heat and Power Limited (SELCHP) and its construction and operation will be in the private sector. The plant will handle 400,000 tonnes of municipal waste a year when fully commissioned and the design emphasis of the £395 million plant is focused on electricity sales to London Electricity in response to the NFFO levy.[101] Interestingly, the public announcement of the scheme met with little local opposition (in stark contrast to similar projects in Hamburg and New York), and there has been support for incineration from the boroughs of Greenwich and Lewisham, though in neighbouring Southwark the administration chose to withdraw from the scheme in order to focus on recycling.[102] Other examples of incineration plants are at an earlier stage of planning and consultation. A proposal has been put forward by the waste management firm Cory Environmental Ltd for an 80–100 megawatt plant in Belvedere, south east London, which if constructed would be among the world's largest incineration plants, handling a third of London's municipal waste stream, equivalent to some 1 million tonnes a year.[103] Unlike the Deptford plant, the Belvedere proposal has met with opposition from local government, businesses and residents concerned about the health and traffic implications of the project.[104]

Studies have suggested that the burning of 40 per cent of the UK municipal waste stream in incineration plants could result in the environmental impact of a reduced consumption of 65 million tonnes of coal equivalent from the production of heat and electricity, coupled with reduced emissions of methane at landfill sites.[105] The privatized electricity utility National Power is currently investigating the feasibility of burning 10 per cent of the UK municipal waste stream and they calculate that some 60 per cent of the income will be derived from waste disposal fees paid by municipal authorities with the remaining 40 per cent from sales of electricity.[106] There is now a policy consensus forming at a national level in the UK for the use of energy from waste as a component of global warming policy. A pledge at the 1990 Labour conference was a commitment to meet the IPCC (Intergovernmental Panel on Climate

Change) target for a 60 per cent reduction in carbon emissions, and this was to be achieved by a variety of measures including the use of electricity generation from municipal waste.[107] Likewise, the Government's 1990 White Paper *This Common Inheritance* sets a target of 1000 MW for renewables by the year 2000, constituting around 2 per cent of total UK energy needs.[108] It seems clear that the waste disposal pattern of the late 1990s and beyond in London is likely to have a growing component of incineration, as suggested by Figure 3.9, particularly as landfill space is exhausted after the year 2000. In this respect the future path of waste management in London will begin to mirror more closely the pattern in other developed economies, where there have been long established economic, political and environmental disincentives to rely on landfill.

Figure 3.9 The location of new incineration plants for municipal waste

Source: Gandy, M (1993), *Recycling and waste: an exploration of contemporary environmental policy*, Avebury, Aldershot

CONCLUSION

The process of 'demunicipalization' in London has been marked by a weakening of any democratic influences on strategic policy making. The choice of options for waste management is being determined increasingly by the profitability of the private sector waste management industry rather than the strategic needs of the community coordinated through elected local government. This process is most clearly seen in the decline of radical waste management in London since the early 1980s and the fragmentation and dissipation of the public sector role in coordinating and providing waste management services.

It is clear that the UK government's target of 25 per cent recycling will not be achieved in London with the current emphasis on market-based waste management policy because of fundamental misconceptions on the part of policy makers as to why the current level of recycling in London is so low in relation to technically achievable levels. In particular, policy makers have failed to appreciate the extent to which a substantial increase in the level of materials recycling would necessitate the better funding and organization of local government and government action to improve the secondary materials markets and control the growth of packaging wastes.

4

New York

After its founding in 1625 the City of New York grew rapidly, reaching a peak of 7.9 million people in 1950, falling slightly to around 7.4 million by the early 1990s. Politically, the City of New York comprises five boroughs: Manhattan, the Bronx, Brooklyn, Queens and Staten Island (Figure 4.1) and the City exhibits a socio-economic diversity and complexity perhaps unrivalled in the western world. New York has one of the highest per capita rates of municipal waste generation recorded, far exceeding the level of waste production in Third World cities such as Calcutta or Manila.[1] The City now faces a 'garbage crisis' derived from the almost total reliance on just one waste disposal facility, the Staten Island landfill, now the largest landfill site in the world, higher than the pyramids of Egypt, and likely to be exhausted sometime in the early twenty first century.[2] The City's proposed solution of building five major incineration plants would still leave some 40 per cent of the municipal waste stream to be disposed of by other means, along with around 3000 tonnes a day of toxic ash residue from waste incineration. The City is now faced not only by escalating costs for its waste management but also by the environmental and health risks of six landfill sites including five former sites, three outmoded municipal incinerators, and a diverse array of smaller hospital and apartment house incinerators.[3] It is against this background that the City has committed itself to a dramatic expansion in the level of recycling and waste reduction.

THE HISTORICAL DEVELOPMENT OF WASTE MANAGEMENT FOR NEW YORK

In the late eighteenth and early nineteenth centuries New York City established municipal control over sewerage and street cleansing but there were jurisdictional disputes between state and municipal governments over the division of responsibilities. The rapid growth of the City in the nineteenth century saw a marked deterioration in public health and by 1870 infant mortality rates were reported to be 65 per cent higher than in 1810. A central focus of concern in the late nineteenth century was street cleansing and the need to dispose of manure produced by the City's 120,000 horses. The late nineteenth century also saw increasing protest

Figure 4.1 The five Boroughs, New York City

Source: Adapted from Goldstein, E A and Izeman, M A (1990) *The New York Environment Book*, Island Press, Washington, DC

against inadequate refuse collection and disposal and the historian Martin Melosi describes the pivotal role of campaigns led by groups such as the Citizens' Committee of Twenty-one and the Ladies' Health Protective Association of New York City.[4] In the 1880s over 70 per cent of New York City's refuse was dumped in the Atlantic Ocean: the refuse barges generally discharged their cargoes at a point midway between the New Jersey and New York shores and about fifteen miles from land. Eventually in 1908, as a result of increasing complaints from the numerous summer resorts on the shores of New York and New Jersey, an investigation into the city's waste disposal was carried out by the Metropolitan Sewerage Commission at the request of the mayor of New York. The report not only found refuse floating in the water over an extensive area; it was also being washed up for fifty miles along the New York shore and for seventy five

miles along the New Jersey shore. It was recommended that if sea dumping was to continue, the refuse should be carried at least a hundred miles
out to sea before being cast overboard. However there was growing dis-
favour for any form of ocean dumping:

> The sewage of the entire population has always been and is now discharged
> into the rivers and other tidal waters in New York City and vicinity with-
> out purification. Until very recent years the refuse from the streets and
> houses was emptied into the harbour also. These customs have, with the
> growth of population and the adoption of higher aesthetic standards, come
> to be regarded as insanitary.[5]

The defeat of the Tammany Hall administration in 1894 brought in the
reformist mayor William L Strong and his pioneering Street-Cleaning
Commissioner George Waring who resolved that waste would no longer
be collected in a mixed state and then simply carried out to the ocean for
dumping. Waring implemented a radical programme of source separation
on the premise that mixed refuse limited the options for disposal, whereas the separation of wastes at source allowed the city to recover some of
the collection costs through the resale and reprocessing of materials.
Under the new arrangements householders were required to separate
their waste into three receptacles. One dustbin was for kitchen waste.
Another was for ashes (the dominant fraction of waste before the decline
of coal for domestic heating). A third contained a variety of reusable materials – chiefly paper, cloth, glass bottles and various metals. Under the
Waring source separation programme those who resisted complying with
the rules were sometimes fined or even arrested but within four years the
initial unpopularity of the scheme had been overcome.[6]

The kitchen waste was collected by the Department of Street Cleaning
in open metal carts and taken to the waterfront where it was conveyed by
barges to a plant at Barren Island at the Rockaway Inlet in lower New
York Bay. The waste was placed in metal retorts and heated with steam,
the grease being separated and chilled, while the residue from the retorts
was pressed and used as the basis for agricultural fertilizer. Both products
were sold by the plant operators, the New York Sanitary Utilization
Company, with contracts for the three boroughs of Manhattan, Brooklyn
and the Bronx.[7] The coal ashes were collected from the houses about once
a week, hauled to the waterfront and then taken by barge to be used for
filling low lying land, the main land reclamation project being based at
Rikers Island in the East River in New York City. The reusable materials
were collected when householders notified the department by placing a
signal card in a window to indicate that there was material to be taken
away. The paper and assorted materials were then taken to central depots
(the first refuse sorting plant in the US was established in the City in 1898)
or to piers along the waterfront where they were sorted and any useful
materials sold by contractors. The City received a small sum of money
from a contractor in return for permission to pick over the refuse. The
picking rights were often sublet to other contractors using immigrant

labour, particularly Italians, and contemporary commentators noted that this work was done 'in a dirty, untidy, and insanitary manner'.[8]

There is evidence that the reuse of materials recovered from New York's waste stream faced difficulties from the state of the secondary materials markets in the late nineteenth century. For example, George Waring describes how the US waste paper trade grew from five million tonnes a year in 1881 to over 22 million tonnes in 1895. In the case of New York City waste paper was collected variously by street collection and regular removal of waste paper from commercial premises, along with separation of paper at garbage dumps and refuse transfer stations. In the 1890s, however, the waste paper market and the rag trade faced major price falls after the introduction of new technologies in paper making employing the increased usage of wood pulp and other new materials.[9] Finally, in 1898, Mayor Strong was defeated by Tammany Hall and much of Waring's waste separation programme was abandoned with a resumption of indiscriminate ocean dumping.[10]

In the early decades of the twentieth century the processing of putrescible kitchen wastes by the steam reduction method fell into disrepute. Martin Melosi describes how

> the reduction facilities polluted the air with the pungent odor of huge quantities of putrefying wastes, which were 'cooked' as part of the process of recovering grease and other products. In addition dark-colored liquids from the compressing process often ran off into the nearby streams.[11]

The odour problem led to the closure of the Brooklyn plant and in 1917 in the face of opposition from local residents a decision was made to transport putrescible wastes to a more remote steam reduction plant on Staten Island to serve the boroughs of Manhattan, the Bronx and Brooklyn. Yet the upsurge in labour costs since World War I and the development of modern fertilizers made from chemical and petroleum feedstocks led to an abandonment of profitable steaming of waste both in New York City and elsewhere in the United States.[12] As for the use of incineration, this did see some developments within the city, with the first combined incineration and electricity generation plant being commissioned in 1905, providing the lighting for the East River Bridge and other local highways.[13]

In 1930, the Commissioner of the newly organized Department of Sanitation announced plans for a city-wide incineration programme involving the construction of 15 plants throughout the five boroughs. This programme was designed to meet the shortfall in waste disposal capacity from the exhaustion of available landfill sites and the growing political restrictions on ocean dumping. The expansion of incineration was given impetus in 1934 when New Jersey coastal cities went to court to force New York City to terminate its ocean dumping. Although only 12 per cent of the City's wastes were still disposed of in this way, there were significant disruptions. The federal Public Works Administration had allocated $4 million for an emergency incinerator construction programme to rehabilitate and expand five existing incinerators and build two new incinerators in Manhattan and the Bronx. Despite the fact that the two new incinerators were ready in time and there was an expansion in the number of landfill sites, most of the extra waste was diverted to Rikers Island where the operations were described as being on the verge of collapse through inadequate equipment in constant need of repair.[14]

One other method of disposal facing decline over the twentieth century was the use of kitchen waste as feed for pigs and poultry. In the 1930s scientific studies in the US established that the use of kitchen waste as feed was an important factor in the infection of pigs with the parasitic nematode *Trichinella spiralis* which could be transmitted to human beings in undercooked meat from the animals. Yet the feeding of pigs with household waste was still very widespread until the early 1950s, with some 630 tons of swill collected per day by the Department of Sanitation, equivalent to around 16 per cent of total incineration capacity in Manhattan.[15] However, a national outbreak of vesicular exanthema in the mid-1950s led to the slaughter of 400,000 pigs and the Public Health Service and state health departments issued regulations forbidding the feeding of uncooked kitchen waste to pigs. Although the cooking of waste did minimise health risks, the practice of using household waste for pig swill declined markedly from the late 1950s for economic reasons and by the early 1960s only 4 per cent of collected food wastes in the US were still fed to pigs.[16]

From the late 1930s onwards there was an increasing reliance on landfill for the disposal of the City's waste. In 1938 Mayor Fiorello LaGuardia built two incinerators equipped with turbine generators to produce significant quantities of electricity but the New York Edison Company refused to buy it, making incineration economically unattractive in comparison with cheaper disposal by landfill. Further factors working against the expansion of incineration were the political power of the Parks Commissioner Robert Moses who utilized municipal waste for the construction of parks and highways, and the continuing difficulty in obtaining suitable sites for the construction of new plants.[17] The Second World War also had an impact with cutbacks on the incinerator construction programme. The shortage of parts and labour led to the closure of many plants, most of which were never re-opened. Thus by 1945 the City's waste management system had been reduced to an inadequate state. The immediate post-war period was marked by the opening of a major new landfill facility at Fresh Kills on Staten Island, and a new capital construction plan for the building of new incinerators along with the rehabilitation of the City's existing eleven plants. Yet only half of the planned incineration programme was carried out, and it was only through the growing reliance on landfill that the City could cope with the growth of the municipal waste stream, which increased by 78 per cent during the decade 1955 to 1965 alone.[18] A further difficulty noted in the 1950s was the air pollution generated by some 7000 apartment-house incinerators, accounting for around 6 per cent of total waste disposal in the city.[19] A 1951 city law actually required all new apartment houses to contain on-site incinerators and by the 1960s around one-third of the City's refuse was being burned in some 17,000 apartment house incinerators and some 22 larger municipal plants.[20] The steady growth of New York's waste stream over the post-war period also coincided with a rapid decline in established recycling activities as demand for recycled products fell in the context of rising labour costs and lower costs for producing products from virgin raw materials. Changes in lifestyle and the proliferation of products which were difficult or expensive to repair contributed to the changing pattern of waste management in the City.[21]

With the rise of environmental concern in the 1960s the problems of waste management began to be conceived as part of a wider environmental crisis and in 1965 Congress passed the Solid Waste Disposal Act. This recognized the growing size and changing characteristics of municipal waste and the difficulties of undertaking economically viable programmes for materials recovery. As late as 1939, 43 per cent of New York City's refuse was ashes but there was a steady increase in the proportion of paper and packaging materials, so that by the 1970s paper was the largest fraction of the waste stream along with increased proportions of glass, metals, and particularly plastics. In 1970 the Solid Waste Disposal Act was modified by Congress to become the Resource Recovery Act emphasizing the need for recycling and the conversion of waste into energy. Responsibility for enforcement of waste legislation passed from the Public Health Service and the Bureau of Mines to the newly created Environmental Protection Agency.[22]

The post-war period has seen a steady decline in the number of landfill sites operating within the City as a result of growing awareness of the effects of landfill leachate and gas emissions in the context of increasingly stringent environmental regulations issued by federal government. Further factors include the City's population growth into vacant areas and the exhaustion of lowland areas suitable for waste disposal. The dozens of sites in the past which afforded some degree of geographical and operational flexibility have given way to the current circumstances whereby the City is almost completely dependent on the Fresh Kills site on Staten Island. Since the creation of the City's Department of Sanitation in 1934, no municipally collected waste has been disposed of outside the City, and even the possibility of sending waste from New York to landfills elsewhere has aroused strong passions: in 1989, for example, the town of Benton in Arkansas rejected a contract to receive waste from New York amid hysteria over the threat of syringes from unmonitored medical wastes and the then Governor of Arkansas, Bill Clinton, supported legislative moves to ban the receipt of out-of-state garbage.[23]

By the early 1980s New York City found itself reliant on one landfill for disposing of most of its waste, as shown in Figure 4.2. With the closure of the Fountain Avenue and Edgemere landfill sites and the continuing operation of only three small incineration plants the City has been left with a radical contraction of past waste disposal options, which once included ocean dumping, a variety of comprehensive recycling initiatives,

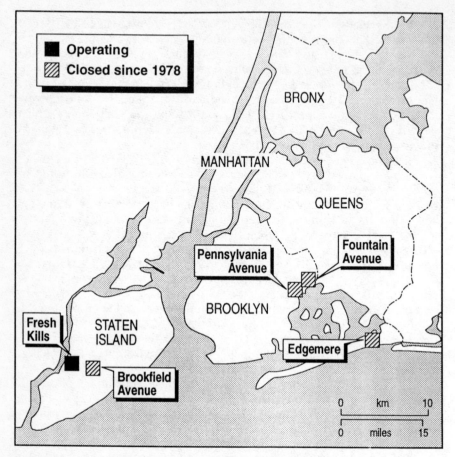

Figure 4.2 New York City landfill sites

Source: Adapted from Goldstein, E A and Izeman, M A (1990) *The New York Environment Book*, Island Press, Washington, DC

over 80 landfill sites (in the early 1930s) and some 22 incinerators.[24] Three main developments lie behind this contraction in available facilities over the post-war period:

1. The increasingly restrictive environmental regulations, marked by the ending of ocean dumping and stringent new controls on the use of incineration and landfill.
2. The exhaustion of landfill capacity in and around the City and the limited potential for land reclamation because of environmental protection of wetland habitats for wildlife.
3. Economic changes associated with the rise of mass consumption, escalating labour costs, declining markets for recycled materials and the fiscal crisis of New York City in the mid-1970s.

THE IMPACT OF THE NEW YORK STATE BOTTLE BILL

The 1970s saw mounting political pressures for controlling the proliferation of non-returnable packaging in the waste stream and a number of US states introduced legislation to impose deposits on beverage containers.[25] For the packaging industry this was seen as a direct and politically motivated infringement on the free market and a threat to profits, as illustrated in the response to the public debate over bottle bills from William F May of the American Can Company:

> We must use every tool available to combat bottle referendums this year in Maine, Massachusetts, Michigan and Colorado where Communists or people with Communist ideas are trying to get these states to go the way of Oregon.[26]

By the early 1980s, however, the political momentum for returnable container legislation in New York State had become unstoppable, and the case in favour of a bottle bill for New York State was succinctly put by Assemblyman G Oliver Koppell of Bronx County:

> Returnable container legislation not only greatly reduces litter, provides enormous savings in the cost of collecting solid waste and conserves significant amounts of energy and natural resources, but actually creates employment without placing any substantial burdens on manufacturers, retailers or consumers.[27]

In 1982 the New York State Legislature passed the Returnable Container Law taking effect from July 1983. In the first full year of operation in 1984 the redemption rate for the 5 cent deposit stood at 71.6 per cent, and the average rate of redemption over the period 1984 to 1991 hovered at around 75 per cent.[28] However, the return rate stands at some 90 per cent in upstate New York compared with little over 50 per cent in New York City, where more affluent residents have less incentive to collect their $5 deposits and there are problems of storage for many smaller apartments.[29] The accumulation of deposits has raised questions over where the money from unredeemed deposits should go, indeed in 1985 Governor Mario M Cuomo proposed measures to New York State Legislature that would enable the State to recoup the unclaimed deposits but these have never been enacted.[30]

One significant consequence of the bottle bill has been the provision of a livelihood for some of the very poorest New Yorkers who can earn around $35 for eight hours work collecting cans and bottles.[31] In its submission to the public hearing on the City's 1992 Waste Management Plan, for example, the Bronx based Homeless Organizations Working Group noted the importance of the R2 B2 buy back centre in the Bronx for providing income for substance addicted and mentally disabled people in the context of a surplus of low-skilled labour in New York City.[32] Although the law insists that retailers must accept up to 240 containers from any one person, in practice many storekeepers tell can collectors to come back later or refuse to redeem deposits, leading to the growth of intermediate organizations such as Two for One which operates through the night to reimburse can collectors in urgent need of cash.[33]

In September 1989 the Governor of New York State, Mario Cuomo, issued an executive order creating a Moreland Act Commission to investigate the effectiveness of returnable container legislation.[34] A key finding of the Commission's report published in 1990 was that the Returnable Container Act had reduced litter by some 72 per cent and had reduced the solid waste stream by over 5 per cent by weight and 8 per cent by volume, thereby making a significant contribution to recycling and waste reduction. The Commission also made a number of important recommendations to improve the operation of the bottle bill:

- The extension of the legislation to include wine and spirits containers.
- The provision of public subsidies to create a viable network of redemption centres and third party recyclers to tackle the low urban redemption rates (attributed to inadequate storage space and lower levels of commitment to recycling).
- The payment of unredeemed deposits to New York State, to be used for the subsidizing of redemption centres, enhancing educational programmes and supporting enforcement of the Act.

Yet the strong endorsement of the bottle bill by the Moreland Commission has provoked opposition in the retail and beverage container industries to the extension of the legislation to other packaging. Examine, for example, the sentiments of the New York State Food Merchants Association:

> In recommending an expansion of the Returnable Beverage Container Law to additional containers, the City proposes to expand the role of food merchants as garbage handlers and to shift the costs of municipal waste management to the private sector, a sector that already pays a disproportionate share of garbage disposal costs – through municipal taxes and through private carting fees. And, since these costs will necessarily and inevitably be passed along to consumers, garbage disposal costs are hidden in an indirect and regressive tax on food. Moreover, the consumer has no direct control over these costs, thus, there is no incentive for behaviour change.[35]

LOCAL LAW 19 AND KERBSIDE RECYCLING

The 1980s saw only limited recycling activity in New York City, mainly private sector collections of paper from commercial premises. A variety of pilot kerbside programmes for residential apartments were also run by environmentalist groups such as the Environmental Action Coalition in order to demonstrate the viability of comprehensive recycling as an alternative to expanding the use of incineration.[36] Finally, in 1987, after some seven years of preparation and extensive lobbying by environmentalist groups, the Department of Environmental Conservation in New York

State released its Solid Waste Management Plan. The plan called for a waste management hierarchy placing recycling and waste reduction ahead of incineration and landfill and laid the basis of the State Solid Waste Management Act of 1988, setting the goal of a 50 per cent reduction in the waste stream by 1997, derived from 8–10 per cent waste prevention at source and 40–42 per cent materials recycling. In addition to the state legislative framework, the City of New York passed its own mandatory recycling law in 1989, Local Law 19, requiring the city to recycle at least 25 per cent of its waste by 1994.

The most ambitious element of the Sanitation's Department's recycling programme is a mandatory kerbside collection programme which from September 1993 extended to all 59 Community Districts in the five boroughs, making it the single largest municipal recycling programme in the world.[37] In most districts there is a weekly collection of newspapers, magazines and corrugated cardboard which are bundled and set out at the kerbside for separate collection in a standard rear-loading compactor truck. A second truck makes a weekly collection of mixed metals, glass and plastics from either blue bins or blue bags placed on the kerbside. And finally, a third truck makes regular collections of unsorted wastes for routine disposal. Some areas within the City such as Park Slope in Brooklyn have now been designated as intensive recycling zones where collections are extended to include textiles, putrescible kitchen and garden wastes, and other materials.

It is the responsibility of building owners and managers to establish a recycling area in their buildings and also to inform tenants how to participate. Participation is enforced by Sanitation Department Police who issue a $25 ticket for initial non-compliance rising to $500 for repeat offenders.[38] In addition to mandatory measures there is an ambitious outreach programme targeted on schools and some 5.5 million pieces of promotional literature were distributed throughout the City in 1991 alone.[39] In fiscal year 1991, the City paid an average of $23 a ton for dealers to accept, process and market paper and an average of $57 for the processing of the mixed recyclables. The newspapers, magazines and corrugated cardboard are taken directly to paper dealers' facilities located within the city, where they are sorted, baled and graded for transport to paper mills, with major exports from Port Newark to timber-poor countries in the Third World. The mixed metals, glass and plastic are taken to a City-owned, privately operated sorting centre in East Harlem or to private processors outside the city, where these materials are cleaned, sorted and compacted for shipping to manufacturers. Figure 4.3 shows how about a quarter of the mixed materials taken to the Harlem sorting facility have to be disposed of, particularly discarded electrical goods, badly contaminated materials and most non PET or HDPE plastic polymers currently misleadingly labelled as recyclable thereby adding to the sorting costs at the plant.[40]

The East Harlem plant provides low-skilled employment for around forty local people and this provision of employment in the context of the dramatic collapse of manufacturing in New York City has emerged as a central rationale for the expansion of recycling.[41] Only about 3400 materi-

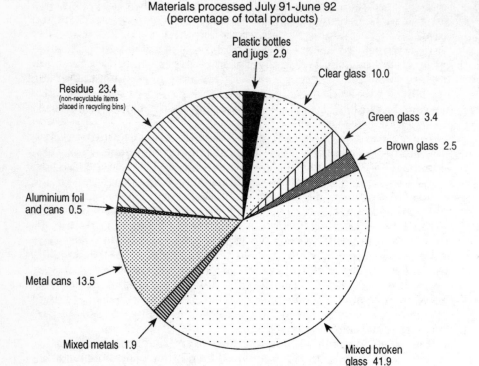

Materials processed July 91-June 92
(percentage of total products)

Plastic bottles
and jugs 2.9

Clear glass 10.0

Residue 23.4
(non-recyclable items
placed in recycling bins)

Green glass 3.4

Brown glass 2.5

Aluminium foil
and cans 0.5

Metal cans 13.5

Mixed metals 1.9

Mixed broken
glass 41.9

Figure 4.3 The East Harlem materials reclamation facility

Source: The City of New York Department of Sanitation

als reprocessing jobs now exist in the City but it is estimated that the City's
recycling and waste management programme will eventually generate
between 44,000 and 60,000 jobs in raw materials processing and related
industries, with higher levels of job creation for increased levels of materials recycling in comparison with incineration.[42] This is significant, since
some 700,000 manufacturing jobs have been lost in the City since 1960
leading to what has been described as a 'dual city' split between an
increasingly specialized high-skill economy (based around financial services and employing predominantly white graduates) and an army of
poorly qualified and casually employed people disproportionately composed of ethnic minorities and other disadvantaged groups.[43] In the early
1990s it is alarming that only 50 per cent of the City's high school entrants
graduated, contributing to a growing mis-match between the levels of
qualifications among New York residents and available job opportunities.
In 1993 the City Comptroller Elizabeth Holzman issued a report calling
for urgent action to ensure that recycling employment is kept within the
City by encouraging the location of waste recycling and reprocessing
firms. This was to be achieved by a variety of measures such as exempting
small businesses from commercial rent tax, improving the quality of

recovered materials, the provision of City-owned vacant land at low cost and lowering energy costs from the State Power Authority to small industrial firms.[44]

The implementation of Local Law 19 through the extension of kerbside collection throughout the City has been slower than anticipated because of a city fiscal crisis, during which recycling temporarily slipped down the political agenda and only just survived the budget battle in the summer of 1990.[45] In the 1990 fiscal crisis the Sanitation Department appeared particularly vulnerable with one possible budget scenario involving a 20 per cent cut in funding. The most vulnerable part of the Department's activities appears to be recycling: the dilemma is that cutbacks in other areas of the Department's work such as numbers of Sanitation Police could lead to a resurgence of problems with the illegal dumping of waste including asbestos and medical wastes, graphically illustrated when hundreds of syringes were washed up on beaches in the summer of 1988.[46]

The City had expected to save some $70 million over four years by delaying the city-wide implementation of kerbside collection.[47] One cost-cutting measure has been the switch from free blue bins to the use of blue bags which must be purchased by householders, but the use of these plastic bags has been criticized, because of their non-recyclability (therefore adding to New York's waste stream) and also because the bags are ripped open by five cent deposit scavengers. Evidence from other cities such as Houston has suggested that free bins encourage higher levels of recycling because residents may view the purchase of blue bags as a hidden tax increase to help pay for the costs of waste management.[48] Indeed, the impact of wider fiscal difficulties on the expansion of recycling underlines the fact that recycling is not pursued as a means to cut costs of waste management in New York City but as a service demanded by the public and by environmental legislation. The costs of kerbside recycling have also greatly exceeded the predicted $65 a tonne, to reach between $198 and $273 per tonne[49] and this has led to direct criticism of the programme in the journal *Public Interest*:

> Recycling enthusiasts believe that the cost of expansion will be paid for by selling recyclables but the City's present Sanitation Commissioner, like his predecessor, finds the costs of an experimental recycling program soaring because of the inefficiencies of collecting and segregation, with no positive relief measures in sight and, in many cases, the marketing potential uncertain.[50]

The lack of progress in the implementation of Local Law 19 led to a legal ruling against the Dinkins Administration by State Supreme Court Justice Irma *v* Santaella in February 1992 after the City was successfully sued by the environmental pressure group Natural Resources Defense Council. Santaella drew attention to the fact that by early 1992 the City was only recycling 1287 tons of waste a day instead of the 2100 tons mandated in the 1989 law to be achieved by March 1992. Also the City had built only one recycling plant, rather than the ten required by law, and no deposit programmes had been established for tyres or batteries as required by

January 1991. Santaella demanded that the City provide details of contract negotiations to build the City's second recycling plant and to prove
that job opportunities for disabled and unskilled New Yorkers had been
created as intended under the 1989 Law. The fact that the City was facing
the worst fiscal crisis since the mid-1970s was deemed irrelevant to its
failure to meet recycling targets.[51] Unlike neighbouring Newark in New
Jersey, however, the financial concerns over recycling and waste management have not led to pressures for privatization and there are concerns
that any loss of City control of operations would lead to greater inefficien-
cies, loss of data, increased involvement of organized crime in waste
haulage and a loss of accountability to elected officials.[52] The slowdown
in the recycling programme during 1991 also led to accusations from the
leading environmentalist Barry Commoner and the Natural Resources
Defense Council that the stalling of recycling was really a conspiracy to
boost incineration.[53] It is suspected that the pro-incineration legacy of the
former Sanitation Commissioner Norman Steisel still exists in terms of
secret plans to construct the proposed five incinerators to replace the Fresh
Kills landfill site.[54] The current level of recycling is estimated to be around
14 per cent in those districts participating in the kerbside collection
scheme. It is therefore still well below the legal target of 25 per cent by
1994 and insufficient to eliminate the need for new waste disposal facilities.[55] A number of environmentalist groups have adopted an
anti-incineration position under all circumstances marked by the *Recycle
First* document submitted for consideration by the Department of
Sanitation in drawing up the 1992 Waste Disposal Plan.[56] One analysis
has suggested that over 50 per cent of the City's waste stream could be
recycled, thereby obviating the need for the construction of any incineration plants.[57] Yet important doubts remain over the long-term costs of
post-consumer recycling; the potential for intensive recycling throughout
the City beyond affluent 'environmentally friendly' areas; and the cumu-
lative impact of increasing recycling activity across North America on the
state of secondary materials markets.

In addition to fiscal constraints from the City budget and weaknesses
within the secondary materials market, the recycling programme has
faced further areas of difficulty. The first of these concerns state and federal level legislation urging the need to reduce waste at source. The City's
1991 Recycling Plan identified four main ways in which waste reduction
could be achieved:

1. The substitution of reusable and durable goods for disposable ones
    such as plastic kitchen utensils.
2. The reusing of materials without significant reprocessing for the
    same purpose for which they were originally intended (particularly
    the use of refillable rather than one-way disposable packaging).
3. The encouragement of home composting for putrescible kitchen and
    garden wastes.
4. The elimination of excess materials by the reduction of unnecessary
    packaging.

Yet the Sanitation Department has been quick to recognise the complexity and difficulty of waste reduction at source in comparison with the recycling of post-consumer waste:

> Reduction of waste is...the least understood option in waste management because it depends on altering buying habits, preferences and manufacturing (and packaging) processes that usually take place outside the locality wishing to reduce or eliminate waste.... Shifting patterns of manufacturing and packaging is difficult and is likely to result in strong opposition from manufacturers, and possibly – at least in the short term – in higher costs and in fewer consumer choices.[58]

A second problem concerns the promotion of recycling in high density, low income areas of the City such as Harlem and the south Bronx. In central Harlem, for example, over 40 per cent of the population receives some kind of public assistance and day-to-day concerns of economic survival clearly take precedence over the sorting of waste for recycling.[59] In high rise apartments, building superintendents usually cannot identify who did not sort their waste and there is consequently less pressure to comply with the law. There are pressing concerns for the safety of residents which may deter people from making use of facilities located in common areas near badly lit stairways and corridors. There is also greater doubt and suspicion from local residents as to whether recycling is simply a revenue raising device by the City and many apartment managers have been reluctant to cooperate with the programme.[60]

A further problem is widespread community opposition, not only to the construction of incineration plants but also towards recycling facilities such as buy back centres and materials recovery facilities because of noise, numbers of truck movements and foul smells. Fears are focused in particular on the impact on property prices from any kind of waste management facility, even garages and storage depots.[61] The location of waste management facilities tends to be concentrated on suitable sites in commercial and industrial areas of the City such as the Port Morris and Hunts Point sections of the Bronx yet this inevitably means that low income black and Hispanic residential areas adjoining these sites will be worst affected, leading to calls for 'fair shares' in the distribution of locally unwanted land uses across the City.[62]

It is too early to judge the overall success of the ambitious city-wide kerbside recycling programme. What is clear, however, is that the 1990s have seen recycling take a key place within the overall waste management strategy for the City, not simply as a palliative for the environmental lobby or as a public relations exercise on the part of municipal government, but as an integral component of public policy with positive social and environmental aims. Yet even if recycling can be made to work effectively in New York there will still be a need to replace the Fresh Kills landfill facility, implying that the ongoing dilemma over the expansion of incineration is certain to be heightened as the City enters the twenty first century.

CONFLICT AND UNCERTAINTY OVER THE FUTURE

In the early 1980s the City's Sanitation Department concluded that the potential short-term contribution of recycling and incineration had been misjudged and the Department consequently looked to an upgrading and extension of operations at the Fresh Kills landfill site as the only viable option. In focusing its attention on the need for greater use of Fresh Kills, the City was suddenly confronted by the scale of the problem since landfill capacity at the site would be imminently exhausted:

> The detailed engineering analyses that were done for the Fresh Kills Operations Plan revealed for the first time the frightening dimensions of the waste disposal problem.[63]

The Fresh Kills landfill site has been in violation of State Law since 1980 for the release of millions of gallons of toxic leachate into nearby water courses.[64] The site also contains unknown quantities of hazardous wastes deposited there since its creation in 1948 and before more stringent environmental regulations in the 1970s. The level of local opposition to the site by Staten Island residents has now reached such proportions that a referendum was passed in 1992 to secede from New York City and become an independent municipality within New York State.[65] The potential for independence from the disposal of New York's waste is complicated, however, by the island's dependence on New York's water supply system.[66] In any case, Fresh Kills is approaching full capacity with estimates of its closure date varying from 2003 to as late as 2011, perhaps sooner if there are restrictions imposed on the out-of-state haulage of commercial waste from the City by private carters.[67]

The vulnerability of the City's waste management system to any curtailment of present facilities can be illustrated by the impact of industrial disputes. The three month tugboat strike of 1979 resulted in an inability by the Sanitation Department to use its barge transport system to deliver waste to the Fresh Kills landfill site on Staten Island, thus cutting off access to what constituted half of New York's waste disposal system. The consequences of this period of disruption were far reaching: additional costs incurred by the city exceeded $4 million; the greater travel distances to and congestion at landfill sites increased the time taken for delivering each load by up to three hours; in order to maintain adequate collection services under these circumstances, street cleaning had to be curtailed and personnel reassigned to waste disposal operations; mandatory overtime had to be instituted to handle the onslaught of activity, with resultant queuing at incinerators and landfills; there was a greater need for repair and maintenance work on the over worked equipment; a marked increase in gasoline consumption took place through the reliance on road haulage; and finally, the streets became dirtier and a source of concern for public health.[68]

Recent efforts to avert the City's impending waste disposal crisis by an expansion of incineration date from the Lindsay Administration

(1966–1973) when the Department of Sanitation began to anticipate the problem of landfill depletion. The Lindsay administration proposed a 'super incinerator program' in the early 1970s and a series of investigations into waste disposal alternatives such as refuse-derived fuel, composting and pyrolysis, none of which were ever implemented. During the Beame Administration (1974–1977) the City's fiscal crisis put an abrupt end to any programme of capital construction for new incineration capacity.[69] As a result of the mid-1970s energy crisis President Carter introduced measures to subsidize non-fossil fuel sources of energy and developed legislation requiring for the first time that utilities must purchase energy produced from alternative sources.[70] This shift in energy policy received strong backing from the packaging lobby, investment bankers and power plant construction firms. Incineration plant constructors included many firms involved in the nuclear industry, now in steady decline since no new nuclear energy plants have been ordered in the US since 1978:

> With considerable turbine manufacturing capacity and no market in sight, the nuclear energy manufacturers found garbage burning a viable alternative, particularly with tax-exempt financing available through Carter's National Energy Acts.[71]

The so-called Wegman Report produced in 1977 by the City's Resource Recovery Task Force and the consultants Leonard S Wegman called for a variety of planned projects including an expansion of incineration. In 1978 the Koch Administration (1978–1989) proposed the construction of the first new incineration plant at Brooklyn Navy Yard and by 1985 an environmental impact statement had been completed and the project approved by the Board of Estimate. The 1992 Waste Management Plan committed the City to proceed with the major new incineration plant at the Brooklyn Navy Yard site, creating consternation at the Dinkins Administration (1990 to 1993) who had been elected on a pro-recycling environmental ticket with crucial support from the Sierra Club and Brooklyn based community groups.[72] The Brooklyn plant is due to begin construction in 1996 but uncertainty remains through the protracted planning and public consultation process. The new plant will have four separate furnaces and a 500 foot high stack, making it the largest incinerator in New York State. The plant will handle 3000 tonnes of waste a day equivalent to some 15 per cent of the wastes collected by the City. If the other four mooted plants in Manhattan, Queens, Bronx and Staten Island were to be constructed, as shown in Figure 4.4, this would account for 60 per cent of the municipal wastes collected by the City and mark a complete reversal in current waste disposal practices away from landfill.

Public debate has focused on the health and environmental impacts of incineration. A first issue concerns the disposal of residual ash. A 1987 survey by the New York State Department of Environmental Conservation found that levels of lead and cadmium in ash from six incinerators exceeded federal guidelines and rendered the ash hazardous wastes.[73] This is significant because the ash from the three existing incineration plants in New York City is being disposed of with household waste at the Fresh Kills landfill site and the lack of any satisfactory arrangements for the disposal of over 900 tons a day of residual ash from a new plant at Brooklyn has been a key reason for delays in starting the project through the need for state and federal permits. Some indication of the degree of opposition can be illustrated by the range of groups who lodged objections to the plan during the period of public consultation: the Clinton Hill – Fort Greene Coalition for Clean Air; the Consumer Policy Institute; the Environmental Action Coalition; the Lower East Side Coalition for a Healthy Environment (with backing from the United

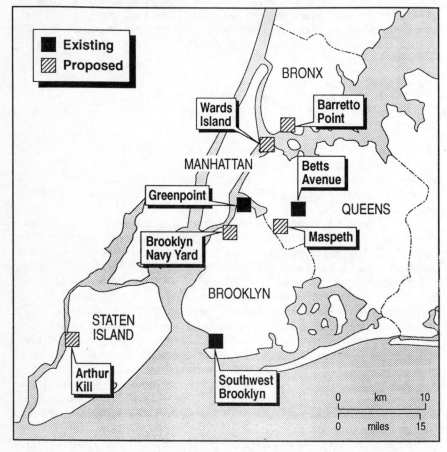

Figure 4.4 New York City's municipal incinerators
(existing and proposed)

Source: Adapted from Goldstein, E A and Izeman, M A (1990) *The New York Environment Book*, Island Press, Washington, DC

Jewish Council and the Physicians for Social Responsibility); the League
of Women Voters and many others.[74] There are a range of specific objections to the Brooklyn Navy Yard project which have been expressed:

- Large-scale incineration in New York City will worsen an already
  serious air quality problem, as suggested by Table 4.1. In the case of
  lead in New York, about 29 per cent of 4974 children screened in the
  summer of 1991 had blood lead levels exceeding the 1978
  standards of no more than 30 µg/dl in children.[75]
- The building of the new incinerator in Brooklyn and the upgrading
  of three existing plants may cost some $1.66 billion or $400 more than
  the Dinkins administration estimated. The operation of the plant may
  also be more expensive than anticipated because of debt servicing,
  technical difficulties, residual ash disposal, and fluctuations in the
  revenue gained from power generation.[76] Studies commissioned by
  environmentalist groups have suggested that incineration will be
  significantly more expensive than recycling, particularly if the
  residual ash were to be managed as hazardous waste.[77]
- The proposed plant will contribute to global warming through the
  emission of carbon dioxide and other gases. New York City is
  vulnerable to global warming through the disruption of drinking

Table 4.1 Projected air pollution emissions from the proposed Brooklyn
Navy Yard incineration plant

Pollutant	*Tons per year*
Nitrogen dioxide	2972
Sulphur dioxide	1189
Hydrogen chloride	537
Carbon monoxide	368
Particulates	161
Sulphuric acid	92
Non-methane hydrocarbons	66
Zinc	28
Formaldehyde	27
Lead	15
Mercury	5

Other pollutants: cadmium; chromium; copper; nickel; arsenic; selenium;
beryllium; fluoride; polyaromatic hydrocarbons; polychlorinated
biphenyls; polychlorinated dibenzo-p-dioxins; polychlorinated dibenzo
furans; tetrachlorinated dibenzo-p-dioxin, and 2,3,7,8-TCDD.

Source: Adapted from Goldstein, E A and Izeman, M A (1990), *The New York
Environment Book*, Island Press, Washington, DC

water supplies by algal blooms in up-state reservoirs and salt water incursion of the River Hudson, in addition to the impact of higher sea levels and greater storm intensity on low-lying parts of the City.[78]

- The City has tried to block the construction of an incinerator in neighbouring New Jersey on grounds that emissions pose a health threat to New York yet this contradicts incinerator safety claims made for the Brooklyn Navy Yard project.[79]
- The expansion of incineration will undermine the impetus for recycling and waste reduction in the City. This will have implications not only for environmental protection but also the generation of employment within the City.[80]

Yet the Department of Sanitation's incineration proposal to proceed with the Brooklyn Navy Yard plant has not been without supporters in the City. The New York Chamber of Commerce & Industry have supported the shift to incineration on the premise that cheaper waste disposal and energy costs are vital for business. Other sources of support have included the construction and retail sector represented by the New York Building Congress Inc. and the Grocery Industry Committee on Solid Waste.[81] The Department of Sanitation has tried to respond to criticisms of incineration by putting the health and environmental risks in a wider perspective. For example, a survey of environmental pollution in the New York–New Jersey region by the US Environmental Protection Agency found that municipal waste incinerators rank as a very low risk in comparison with the medium risk posed by landfill sites and the very high risk from motor vehicle exhaust fumes.[82] The projected lead emissions from the proposed Brooklyn Navy Yard plant of 14.5 tons a year have been singled out as a threat to children's health, yet it is calculated that the energy generated by the plant will displace one million barrels of oil a year used by Con Edison, which would emit some seven times more lead for an equivalent generation of energy.[83] In particular the Department of Sanitation argues that new incineration technologies are safe; that incineration reduces the volume of waste thereby significantly extending the life span of the Fresh Kills landfill; and that burning waste is cheaper than a complete reliance on recycling and waste reduction.

CONCLUSION

The tough legislative framework at State and City level has clearly been instrumental in pushing the most comprehensive recycling programme for New York City seen during peace time since the days of commissioner George Waring in the 1890s. Yet even if the state's source reduction and recycling goals are met *and* the Brooklyn Navy Yard incineration plant is constructed the City will still face a crisis in its waste management with the closure of the Fresh Kills landfill site some time in the early twenty first century. Yet the available options for the future will not be determined by policy debate within New York alone: wider national developments concerning the legislative control of packaging and toxic and unrecyclable products will play a major role in determining the future path of waste management in the US. Furthermore, the pressing demands on the fiscal resources of New York City will continue to determine the limits to comprehensive recycling programmes incurring greater expense than an expanded reliance on waste disposal by incineration:

> Since every dollar from the City's highly constrained budget that is spent on waste management is not available for such other crucial services as police protection or patient care, there is a compelling need for an integrated waste management system that... minimizes these overall costs to the greatest extent possible.[84]

5

Hamburg

The city of Hamburg lies on the mouth of the River Elbe in northern Germany and is the second largest city in the federal republic. In 1991 the population of Hamburg stood at some 1,662,000 compared with a peak of around 1,850,000 in 1965. Administratively the city has special status, as a federal *Land* in its own right, and is therefore represented with the fifteen other *Länder* in the powerful *Bundesrat* second chamber, as are the cities of Bremen and Berlin. Figure 5.1 shows how the city of Hamburg is subdivided into 104 *Stadtteile* and seven *Bezirke*, similar in size but with a weaker service delivery role than the London Boroughs. Only the Bezirk Bergedorf and Harburg in the south of the city assume more wide-ranging responsibilities for the provision of local government services. This is an historical anomaly arising from local government reorganization in 1937, extending the city boundaries into more autonomous outlying areas.[1] Hamburg is currently faced with a deteriorating waste management situation worse than either London or New York, derived from the absence of any landfill opportunities within the city boundaries coupled with escalating political and economic restrictions on landfill outside the city. The position has been exacerbated by German reunification and the restrictions on cheap waste disposal opportunities in the former East Germany. Furthermore, the largest of the city's three municipal waste incinerators is to be decommissioned in 1994 and the costs of waste management are steadily escalating to emerge as one of the most contentious areas of public policy for the city.

THE HISTORICAL DEVELOPMENT OF WASTE MANAGEMENT FOR HAMBURG

By 1900 the population of Hamburg approached 800,000, and the city had become the fourth most important seaport in the world, after London, New York and Liverpool. In the nineteenth century the city faced a public health crisis, with no less than fourteen cholera epidemics occurring between 1831 and 1873. In 1892 there was a major cholera epidemic traced to polluted drinking water from the River Elbe, in which over 8000 people died. As a result the city administration was transformed, constructing a water purification plant and setting up a municipal sanitary institute in 1893. The progress towards better standards in public health for Hamburg

Figure 5.1 The administrative boundaries for the city of Hamburg

Source: Gandy, M (1993), *Recycling and waste: an exploration of contemporary environmental policy*, Avebury, Aldershot

appears to have been very swift. It was noted as early as 1909, in a major review of street cleansing in twenty different world cities, that Hamburg had already developed 'a peculiar and very satisfactory form of municipal government...the streets are, many of them, handsomely built, well paved, and very well kept'.[2]

In the early part of this century the city of Hamburg was served by one large incinerator at Ballerdeich, constructed between 1893 and 1896. This Horsfall type destructor, with its one hundred workers, was said to be among the best and largest of the time, incorporating a number of novel technical and operational features:

• The works were situated on the outskirts of the city and appeared 'to produce little offence on account of odour'.[3]

- There were weighbridge facilities, collecting data concerning the quantity of the refuse which was handled by the works.
- A high labour efficiency was reported for the stoking process.
- A high temperature incineration process was used to reduce emissions, with the remaining ash and clinker[4] cooled and sifted to remove metals, which were then sold.

The dust and clinker was used as raw materials for building purposes, and it was noted that, 'at times the work of building is so active that the production of ground clinker scarcely meets half of the demand'.[5] The steam raised from the four boilers was used both for the running of the machinery necessary for the operation of the plant, and also for the production of electricity, which was transported and consumed in other parts of the city.[6] The emergence and predominance of incineration as the solution to municipal waste management in Hamburg in the early part of the twentieth century suggests that Hamburg, London and New York City initially followed a similar path of development. In the 1930s, however, their development began to diverge, with the emphasis in London and New York moving increasingly towards the use of cheap and accessible landfill rather than incineration. Yet Hamburg was never able to develop a virtual reliance on landfill because of administrative and political restrictions imposed by the neighbouring local government administrations to accepting large quantities of waste, resulting in the need to focus waste disposal within the city boundaries.[7]

The present spatial and organizational structure of waste management in Hamburg was established in 1949 with the creation of a new city cleansing department, the *Stadtreinigung,* handling the collection and disposal of all municipal waste. From the mid-1950s onwards the city's waste stream expanded rapidly as an outcome of the German *Wirtschaftswunder* of rapid economic growth and the well established networks of materials recovery for reasons of economic survival fell into a decline.[8] In the period 1949 to 1985, the volume of household waste grew by 86 per cent to some 5.4 million cubic metres, and over the thirty year period 1952 to 1982, the density of the waste fell by 71 per cent, primarily reflecting a decline in the use of coal for heating purposes and the increasing quantity of packaging materials. In response to the growth of the waste stream, there was an extensive programme of investment in new incineration capacity in the 1960s and 1970s, and the contemporary waste management infrastructure for incineration was fully operational by 1979.[9]

As in the US and the UK, there have been political and economic pressures since the mid-1970s to control the rising costs of municipal waste management. The concern over this issue was resolved, at least temporarily, in Hamburg by the splitting of the operational side of waste management off from the *Baubehörde* (Public Works Department) in 1988 to form a public corporation owned by the city of Hamburg, called the *Landesbetrieb Hamburger Stadtreinigung* (LB-HSR), handling both the collection and the disposal of waste. The LB-HSR remains under political control of the elected city Senate, which has a 100 per cent shareholding and strategic policy decisions are made by elected politicians in the Senate.[10] The core of German waste management policy is contained within the 1972 Waste Management Act. Under the fourth amendment to the 1972 Waste Management Act, which came into force in November 1986, there is now a focus on recycling and waste reduction as an integral element in waste management.[11] Other important legislative developments include the *TA Abfall* technical regulations for waste management initiated in 1984, concerning higher environmental standards in waste disposal with a particular emphasis on incineration emissions and the more recent legislative controls on packaging introduced in June 1991.

THE CONTEMPORARY PATTERN OF MUNICIPAL WASTE MANAGEMENT

A charge called the *Müllgebühr* is levied for the collection and disposal of household waste by the City, and this charge rose by 17 per cent during 1990 alone.[12] It should be noted that of the 860,000 households in Hamburg, only 140,000 are charged *Müllgebühr* separately and individually by the city, since the charge is normally incorporated into rent in the non owner-occupied housing tenures. Hamburg is currently served by four incineration plants and four landfill sites, as shown in Figure 5.2. During the late 1980s, the incineration option for Hamburg faced increasing difficulties. First, there was pressure from the neighbouring *Land* of Schleswig-Holstein to have exclusive use of the Stapelfeld incineration plant because the *Land* had decided to stop sending its waste to landfill in the former East Germany. Second, the large Borsigstraße incineration plant commissioned in 1967 will have ended its working life by 1994, substantially reducing Hamburg's incineration capacity. Finally, the Hamburg Senate had decided in 1986 to phase out the use of the important Schönberg landfill site in the former East Germany, a decision which has been given added impetus by German reunification and concern over environmental standards in eastern Germany.[13]

The use of landfill has been faced with intense political restrictions from the powerful system of local government and strong local opposition to new sites since the early 1970s; the last remaining site within the city area having been exhausted in October 1986. As a result of these constraints, there has been intense political pressure on Hamburg to solve its waste management difficulties within the city boundaries. The only new landfill sites in the city area under consideration in the 1980s were on the edge of the city at Duvenstedter Brook, Hummelsbüttel, and Rahlstedt, and all presented serious difficulties in terms of land use planning and potential opposition from local residents.[14] The neighbouring *Länder* of Schleswig-Holstein and Lower Saxony have shown increasing resistance to taking any of Hamburg's waste for landfill, which explains the emergence of the former East Germany as the main recipient of waste, where there were lower levels of public resistance to landfill and the need for hard currency. In the period 1986–1990 between 213,800 and 538,400

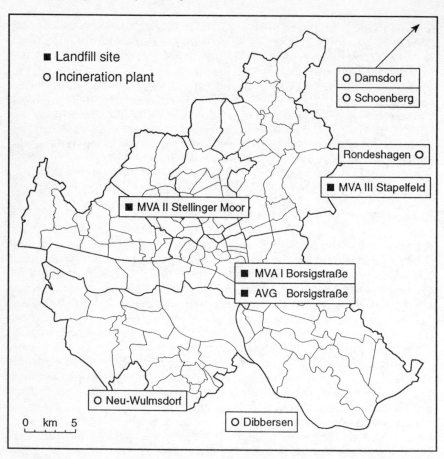

Figure 5.2 The location of incineration plants and landfill sites for
Hamburg's municipal waste

Source: Gandy, M (1993), *Recycling and waste: an exploration of contemporary
environmental policy*, Avebury, Aldershot

tonnes of waste that could not be burnt in Hamburg, were transported
annually 120km by road to be landfilled at Schönberg in the Mecklenberg
region of eastern Germany, along with 230,000 tonnes of sewage sludge
and 90,000 tonnes of hazardous waste.

Incineration has been consistently the main waste disposal option for
the city for a combination of political and organizational reasons stem-
ming from the administrative structure of local government. The powerful
federal regional government structure imposes greater political and economic constraints on the use of landfill for waste disposal, along with the
absence of any semi-rural outlying parts of the city comparable with
Staten Island for New York City. Hamburg is faced with finding, as far as
possible, a satisfactory solution to its waste disposal problem within its
own administrative boundaries. This in turn, accounts for the relatively

high cost of waste management in Hamburg in comparison with London and New York and the resulting political difficulties in finding a cost-effective and environmentally acceptable waste management strategy for the city.

The development of recycling in Hamburg has been associated with a number of distinct phases of interest: the historical promotion of recycling for economic reasons, illustrated in the exchange of waste materials between different firms in the Chamber of Commerce; the recycling of materials for social and charitable purposes, as in Red Cross collections of paper and textiles; the emergence of environmentalist recycling from the urban based environmental movement since the 1970s; and finally, the emergence of recycling in the 1980s as an integral element in municipal waste management in response to declining landfill availability and strong public opposition to incineration. In the contemporary context, the recycling of household waste in Hamburg is seen as an environmental service, and not as an additional source of income or a means of reducing the costs of waste management for the Hamburg Senate.[15] Indeed, the limited income that was to be made from the recycling of glass, and to a lesser extent paper, accrued almost wholly to the private sector, who undertook operations under contract to the city.[16] The recycling rate for municipal waste in Hamburg for the period 1988–1990 is estimated by the city to be around 13 per cent[17] and is therefore comparable with districts participating in the kerbside collection scheme within New York but significantly higher than the level of recycling in London. In the case of Bezirk Harburg, widely considered to have the highest rate of recycling (around 20 per cent) in the city, the recycling programme included a number of different initiatives:

- a 'bring' system of on-street collection facilities for glass and paper;
- a kerbside 'collect' system for paper;
- the collection of a wide range of materials at a major recycling centre with four full-time staff;
- collections of textiles, paper and glass at schools and offices; and
- a dual-bin scheme for the separate collection of putrescible kitchen and garden waste for the centralized production of compost.[18]

Paper is collected mostly under contract by two private firms. On-street collection facilities across the city are used for lower grade materials, particularly newspapers, most of which are taken to a mill 50km outside the city. There have been considerable difficulties since the 1987 collapse in the price of waste paper, forcing the city to take over many loss making waste paper collections, to avoid a political backlash from the electorate.[19] As in the case of glass, there have been disagreements between the city and the private sector over the interpretation of the contract obligations concerning responsibility for the maintenance and upkeep of the collection sites, which could incur major labour costs, potentially greater than the current value of the waste paper in the collection facilities. The glass from bottle banks is collected mainly by two firms under contract to the city. They provide the collection facilities and take the glass cullet to a processing plant 150km outside the city for 60 DM a tonne.[20] As in the UK, there is concern over the possible saturation of the market for green and mixed glass at a national level since 50 per cent of domestic glass production is for clear glass.[21] The main local difficulties reported for glass recycling were similar to those in London, being principally the impact of noise and broken glass in the vicinity of the collection points, and contamination of white glass with coloured glass, necessitating expensive further sorting of the glass cullet.[22] Since 1990, a growing difficulty between the city and the private sector has been the interpretation of the contractual obligation for site maintenance and cleansing, and also over demands from the private sector for a special subsidy to help cover their losses created by increasing price instability in the glass market.[23]

Recycling in Hamburg is marked by the predominance of the 'bring' system as in London, with the use of kerbside collections restricted to some voluntary sector initiatives and the use of the pilot dual-bin system for putrescible kitchen and garden waste in Bezirk Harburg. In addition to the network of on-street collection facilities there is a well developed system of combined recycling centres and civic amenity type facilities across the city where the recovery of CFCs from refrigerators and the renovation of discarded consumer durables is undertaken as part of a variety of state supported employment creation schemes. Figure 5.3 shows the catchment areas of the twelve recycling centres run by the city. Each of these centres has at least two full-time qualified staff, and they handle the whole range of recyclable materials brought by the public. In addition to the recovery of materials, the centres perform an important advisory and educational role, particularly for toxic wastes. Hamburg also provides a number of 'recycling buses', making regular stops across the city, to which the public can bring any toxic household items.[24]

THE RECYCLING OF PUTRESCIBLE KITCHEN AND GARDEN WASTE

The main impetus for the promotion of composting has come from the Hamburg Senate, who have demanded a radical expansion of this option as part of a wider strategy to reduce the reliance on landfill and incineration and also to reduce methane emissions and other environmental impacts of putrescible wastes. The expansion of composting can be seen as a policy initiative arising from the elected politicians in the Senate, rather than a policy promoted for conventional technical or economic reasons, as part of the routine waste management responsibilities of the city. The 1989 Waste Management Plan identified the composting of putrescible waste as a key component of a comprehensive recycling programme capable of raising levels of recycling towards a 30 per cent target for the city.[25] The significance of composting is that some 29 per cent of the household waste in Hamburg is derived from putrescibles, yet there is a need to devise a system of composting for households without gardens. It

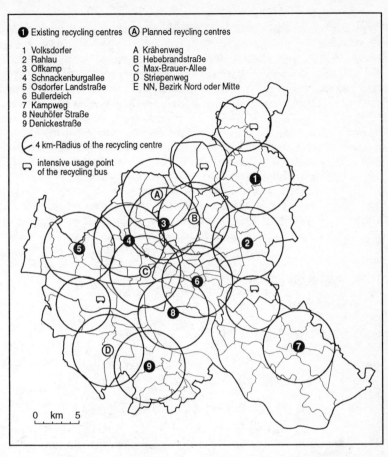

Figure 5.3 Recycling centres in Hamburg

Source: Gandy, M (1993), *Recycling and waste: an exploration of contemporary environmental policy*, Avebury, Aldershot

was envisaged in the 1989 Waste Management Plan that higher rates of composting would be met by two main means: first, the city's promotion of composting by households themselves with access to a garden, as part of the *Eigenkompostierung* programme; and secondly, the development of a dual-bin system for the separate collection and centralized production of compost from households without gardens. The 1989 Waste Management Plan expected that by 1995, up to 80,000 tonnes of waste per year would be handled by four composting facilities, located in Bezirk Altona, Bergedorf and Harburg, and just outside the city boundary in the neighbouring *Land* of Schleswig-Holstein.

The first element in the Hamburg composting strategy, the production of compost by households with gardens in the *Eigenkompostierung* programme, has been promoted since 1985 using the incentive of free provision of composting units which would have normally retailed at 20

DM each.[26] The objective is to enable households to need a smaller bin for their routine waste collections and hence incur a lower *Müllgebühr* charge, so as to stimulate recycling activity. Figure 5.4 shows how the potential for households to carry out their own composting is closely related to housing structure and the number of households with gardens. In theory, the incentive scheme was possible because the *Müllgebühr* charge is made on the basis of bin size. In practice, however, this objective has proved problematic, since the 17 per cent increase in the *Müllgebühr* in 1990 incorporated a 2 per cent rise for larger bins, and a 40 per cent rise for smaller bins. This had followed demands from the city's audit commission, the *Rechnungshof*, that the existing anomaly whereby the larger bins were effectively subsidizing the smaller bins be redressed.[27]

The second element in the composting programme was the separate collection of putrescibles for the centralized production of compost. This

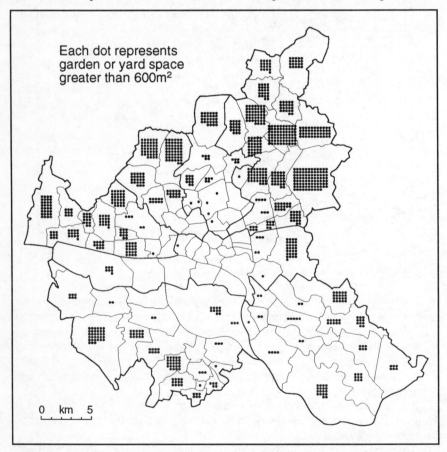

Figure 5.4 The extent of potential composting by households as part of the *Eigenkompostierung* programme

Source: Gandy, M (1993), *Recycling and waste: an exploration of contemporary environmental policy*, Avebury, Aldershot

aspect of policy is being carried out in Bezirk Harburg and Bergedorf in the south of Hamburg, involving households from a wide range of housing types. The scheme uses a dual-bin system for wet (ie putrescible) and dry waste, the putrescibles being collected either weekly or fortnightly and taken to the centralized composting facility nearby. The aerobic composting process takes six to eight months, with labour costs reduced through the use of the long-term unemployed at the plant as part of a state subsidized job creation scheme. The scheme was a free service in Bezirk Harburg, with no increase in the *Müllgebühr* until October 1990, after which a 30 per cent rise in the *Müllgebühr* charge was introduced to cover the extra cost of the scheme. The project has also been run on a voluntary basis, contrasting with mandatory schemes in operation in some other German cities.[28]

A pilot dual-bin system was tested in Bezirk Harburg in the late 1980s. The analysis of the Bezirk Harburg project found that the main contamination problem was from the presence of non-putrescible waste, such as metals and plastics, the proportion varying from 0.2 per cent to 7 per cent depending on the housing type. The highest levels of non-putrescible contamination were found in shared bins serving high rise public sector housing and this necessitated the expensive cleaning of the compost with sieves before sale, although unsightly fragments of metal and plastic always remained.[29] The levels of public participation varied from 55 to 80 per cent of the households in the catchment study, depending on the housing type, with the highest levels of participation in larger detached and semi-detached homes. The explanation for the lower participation rates and higher levels of compost contamination in high rise housing is thought to rest on three main factors:

1. The relative social anonymity of communal bins, as a psychological encouragement to contamination with non-putrescible waste and lack of care in waste separation.
2. Householders without gardens were less acquainted with gardening and the potential use of organic wastes.
3. The influence of higher levels of socio-economic disadvantage on participation in environmental policy.[30]

Turkish people form the main ethnic minority in Hamburg, concentrated in high rise accommodation, and some households may have had difficulty reading the German public information leaflets. In contrast, in New York all recycling information is distributed in English, Spanish, Korean and Chinese.[31]

The main economic obstacle to centralized composting appears to be the higher cost compared with other waste disposal options, as suggested by Table 5.1. Calculations for the Bezirk Harburg project in 1990 put the cost at some 430 DM per tonne, compared with between 242 and 290 DM for waste collection and disposal by incineration and around 240 DM for landfill.[32] One possible way of reducing the cost to 380 DM per tonne was through fortnightly rather than weekly collections to cut labour costs, but

this presented new sets of problems in terms of the increased smell, especially in summer, and the associated lower levels of public participation,
which have also been observed in other pilot projects elsewhere in
Germany.[33] The problem of smell in Hamburg was also greater for the
high rise collections, where the putrescibles were dominated by wetter
kitchen waste decomposing faster than garden derived waste.[34]

A further difficulty is the space requirements of the composting facilities in competition with other more profitable urban land uses. There is
also a need to site the facility at least 500 metres from a residential area,
because of the local environmental complaints created by the smell of
putrescibles in the early stages of composting, though there are plans in
future to utilize anaerobic methods which, while more expensive, would
be quicker, require less space, and produce a less pungent smell, with the
added advantage of being integrated into bio-gas energy recovery
schemes. Despite the low levels of heavy metal contamination for the
compost,[35] the strength of the market for the final product in competition
with commercially marketed peat (a non-renewable resource) is not yet
fully known, and is a key factor influencing the overall economics of an
expansion in composting.[36]

Table 5.1 Comparative costs of waste management options in Hamburg

Waste management options 1988–9	Cost per tonne (DM)
Provision, emptying and transport of glass and paper from on-street collection facilities (Average price under contract)	60
Waste collection and disposal by landfill (Schoenberg site in the former DDR)	240
Waste collection and disposal by incineration (Average cost at the city's three plants)	290
Fortnightly collection of putrescible waste and centralized production of compost (Bezirk Harburg pilot project for dual-bin system)	380
Weekly collection of putrescible waste and centralized production of compost (Bezirk Harburg pilot project for dual-bin system)	430
Bring system of plastics collection including sorting and transport costs (Bezirk Bergedorf pilot project)	500
Kerbside collection of mixed plastics including sorting and transport costs (Bezirk Bergedorf pilot project)	1000

Source: Gandy, M (1993), *Recycling and waste: an exploration of contemporary
environmental policy*, Avebury, Aldershot.

The example of composting in Hamburg illustrates that the achievement of very high rates of urban recycling would necessarily involve the use of putrescible waste from households without gardens. The results of the Bezirk Harburg project suggest that composting of household waste is more difficult than is widely appreciated in the literature,[37] and that considerable uncertainties over the cost and long-term public participation in such schemes remain. Analysis of the projected costs of different recycling strategies suggests that the use of a comprehensive dual-bin system as part of a programme to raise levels of recycling to 30 per cent would raise the costs of waste management by between 89 and 144 DM per tonne, depending on the combination of different methods of materials recovery and whether the separate collections of putrescibles were weekly or fortnightly.[38] The implications are that the extension of recycling to the environmentally significant waste fraction of putrescibles (in terms of methane emissions and the protection of wetland ecosystems from peat extraction, along with the overall goal of waste reduction) is significantly more expensive than routine waste disposal by landfill or incineration, and may have socially regressive and politically unpopular consequences through a forced increase in the flat rate *Müllgebühr* charge to individual households for the collection and disposal of their waste.[39]

THE POLITICS OF WASTE MANAGEMENT IN HAMBURG

Hamburg is faced with a potential crisis in its waste management from four main sources:

1. Intense political pressure to increase the level of recycling, led principally by the Greens in the Hamburg Senate.
2. A high and rising *Müllgebühr* charge to individual households for waste management services.
3. The decision to cease landfill operations at the major Schönberg site in the former East Germany.
4. The imminent decommissioning of the oldest and largest of the city's three incineration plants at Borsigstraße in 1994.

The scale of the contemporary difficulties facing Hamburg can be illustrated by the fact that the Schönberg landfill site and the Borsigstraße incineration plant currently handle, between them, over half of Hamburg's municipal waste stream. Hamburg's difficulties are further heightened by the inability of the city to finance new incineration capacity itself. The cost of waste management in Hamburg has become increasingly contentious in the Senate in the face of growing fiscal difficulties, and this situation has created an intense political debate within the ruling SPD administration over the appropriate role for the private sector in waste management.[40] The left of the ruling SPD administration had pushed for a major waste reduction and recycling strategy coupled with new landfill sites, in order to reduce the need for incineration. However, the newly ascendant right wing of the SPD along with their centrist FDP coalition partners has, since 1987, largely abandoned this approach because of the cost implications for waste management, and has shifted the policy emphasis towards the goal of reduced waste management expenditure by the city.[41]

The political difficulties facing the SPD controlled Senate's waste management policies have been heightened by fluctuations in their share of the vote since the early seventies, as indicated by Table 5.2, forcing them into coalition with the centrist FDP in order to gain an absolute majority in the Senate. In contrast to the SPD, the FDP have wanted to see most waste management handled in the private sector in order to cut costs.[42] The FDP proposed the experimental privatization of waste collection in Bezirk Bergedorf and Harburg as the first step towards complete privatization of waste management, called for by the centre-right CDU opposition in the Senate.[43] The Greens, represented by the radical *Fundis* wing in Hamburg, are opposed to any role for the private sector in waste management, and have advocated the elimination of incineration as an option, coupled with the maximum technically achievable increase in recycling and waste reduction at source.[44] For parts of the environmental movement, including the Hamburg Greens, the rejection of incineration has assumed a symbolic position paralleled by the outright opposition to nuclear power.[45] After the 1987 elections, there were negotiations in Hamburg over the possibility of forming a coalition between the Greens

Table 5.2 Elections to the Hamburg Senate, 1974–91 (% of vote)

	CDU	SPD	FDP	Greens
1974	40.6	44.9	10.9	–
1978	37.6	*51.5	4.8	**4.5
1982a	43.2	42.8	4.8	7.7
1982b	38.6	*51.3	2.6	6.8
1986	41.9	41.8	4.8	10.4
1987	39.3	46.5	6.0	7.1
1991	35.1	*48.0	5.4	7.2
1993	25.1	40.4	4.2	13.5

*   The SPD gained an overall majority of the seats in the Senate
**   This figure includes 3.5% for the BLW (Rainbow Slate) and 1.0% for the GLU (Environmentalist Green Slate)

The 1993 elections also saw other parties attracting a significant share of the vote: STATT Party (5.6%) and Republikaner Party (4.8%)

Source: Mintzel, A and Oberreuter, H eds (1990), *Parteien in der Bundesrepublik Deutschland*, Bundeszentrale für politische Bildung and the German Embassy, London.

and the SPD, as was the case in West Berlin, to create a 'red-green' coalition.[46] However, the policy gulf between the *Fundis* dominated Greens and the traditional centrist orientated SPD proved unbridgeable, and the SPD eventually formed an administration with the FDP, as it has always done in the past, on failing to gain an absolute majority in the Senate.[47] The September 1993 elections to the Hamburg Senate saw a major upset, with both the main SPD and CDU parties' share of the vote slumping to a post-war low and the centrist FDP falling below the 5 per cent threshold for representation in the Senate, while the Green Alternative List substantially increased their share to over 13 per cent. This has put the prospects for a red/green coalition firmly back on the agenda and may yet have important implications for the expansion of the city's recycling programme.[48]

Within the 1987 to 1991 SPD/FDP coalition, the FDP agreed to a compromise over the privatization of waste management, insisting only on the eventual privatization of individual services provided by the city, such as bulky refuse collection and street cleansing, and the continued dominance of the private sector in the provision and operation of on-street collection facilities for glass and paper.[49] The opposition CDU in the Hamburg Senate has come down in favour of the increased use of incineration, and in the early 1980s joined with the Greens in opposition to a proposed landfill facility in a nature reserve within the city boundaries. The CDU arguments against landfill rested on three main elements: its contribution to global warming from methane emissions;[50] the expense of future decontamination and land reclamation of former landfill sites at Tegelweg and especially at the major Georgeswerder site;[51] and finally, the belief that Hamburg should handle its own waste within the city boundaries without relying on complex and expensive negotiations with the neighbouring *Länder* of Schleswig-Holstein and Lower Saxony.[52]

The proposed solution to Hamburg's waste disposal crisis, following heated debate within the SPD city administration, was to give the go-ahead to a joint private venture between the Hamburg based electricity utility, Hamburger Electricitätswerke (HEW), and the national VEBA Kraftwerke Ruhr (VKR), to build one of the largest incineration plants in Germany on derelict land on the Borsigstraße site in the Hamburg docks. During the 1980s there has been widespread diversification within the German energy industry, as state owned utilities such as AVG in Nord-Rhine Westfalen, and other private sector concerns including Babcock, Preußen Electra, Siemens and VEBA move out of nuclear energy (which has been in decline since the late 1970s) and into the incineration of municipal waste.[53] The new Borsigstraße incineration plant is planned to be operational from 1994, and the city will be under contract to supply it with at least 40 per cent of Hamburg's municipal waste stream per year. In addition to receiving a guaranteed income from waste disposal at the site, the operators will also make substantial profit from the utilization of waste heat and the sales of electricity generated.[54] Further impetus for expanding incineration in the city is derived from the rising costs of land-

fill which by 1993 clearly exceeded the costs of disposal in existing incineration plants by around 20 DM per tonne.[55] In Germany as a whole, the
trend towards greater use of incineration seems to be gathering pace. The
new technical regulations for municipal waste passed in 1993 will tighten
environmental controls on landfill to such an extent that it is predicted
that a further 36 incineration plants will be built in Germany by the early
twenty first century.[56]

The 1989 Waste Management Plan drawn up by the city's planning
department and approved by the Hamburg Senate brought to a head the
political conflict over the future of waste management in Hamburg, and
in September 1989 there was a major street demonstration against the
plan. The main groups involved in the demonstration were workers from
the city cleansing department represented by their union Öffentlicher
Dienste, Transport und Verkehr (ÖTV), campaigning against further privatization plans, along with numerous environmentalist and citizens'
action groups campaigning against the construction of the new incineration plant planned for Borsigstraße, and the associated risks from the
emission of dioxins, heavy metals and other pollutants.[57] The opposition
to the 1989 Plan was marked by the production of an alternative plan, the
Müllkonzept, drawn up by the Hamburg Green Party, environmentalist
groups and a Hamburg based environmental consultancy Ökopol GmbH.
The alternative plan called for the cancellation of the Borsigstraße project,
enabled by a combination of greatly increased recycling and waste reduction at source; the adaptation of existing incineration capacity at Stapelfelt
and Stellingermoor to take more waste; and the use of new landfill sites
outside Hamburg, based on full public planning and consultation proce-
dures.[58]

The Hamburg Greens have campaigned for the separate collection of
putrescible waste throughout the city as an extension of the Bezirk
Harburg pilot project, except where this is impracticable, or where households with gardens can carry out their own composting as part of the
Eigenkompostierung programme.[59] The alternative plan claimed that the
Borsigstraße incineration project, approved by the Senate, would be the
most expensive option in comparison with increased recycling and landfill, though this is disputed in other analyses.[60] Most importantly,
however, its construction was seen as a disincentive to expand recycling
and waste reduction because of the contractual controls which would be
instituted over the waste management activities of the city. In particular,
the alternative plan drew attention to the contractual obligation on the
city to provide 40 per cent of the municipal waste stream in Hamburg for
the Borsigstraße incineration plant. The key concern is that the successful
operation of incineration plants does not rely on waste reduction, and a
larger waste stream simply presents greater opportunities to build more
incineration capacity in the future:

> Its construction would be a signal to industry that changes in the produc-
> tion process to increase waste reduction and recycling are not essential.[61]

Theoretically, however, the Hamburg Senate would still have some control over the plant, through their majority share holding in the city's electricity company Hamburger Elektricitäts Werke (the partner with VKR in the joint construction venture). Yet increasingly, local authorities in Germany are selling off their share holdings in such utilities, both for ideological reasons, in terms of extending privatization, and also as an additional source of income. Indeed, there was speculation during 1990 that the SPD administration might sell off the city's cleansing department either as a whole, or split up into different areas spatially,[62] and an offer had already been received from a French waste management firm.[63] If carried out, this would in effect be a complete privatization of all operational aspects of waste management in the city, as is currently taking place disposal groups formed after the abolition of the Greater London Council are opting for outright privatization rather than the formation of LAWDCs under the 1990 Environmental Protection Act.

Hamburg is faced with a complex set of political and economic dilemmas in resolving its future arrangements for waste management. This situation has arisen for a combination of logistical, economic and political reasons, mainly beyond the effective control of the Hamburg Senate. The increasing dominance of incineration in the future could, paradoxically, lead to a lessening of the pressure to expand recycling, despite the stated objectives to expand recycling in the 1989 Waste Management Plan. This and other fears make the new Borsigstraße incineration plant the central focus of conflict over the future of waste management for the city, between the two main strategies: the 1989 Waste Management Plan put forward by the Senate, involving increased incineration in the private sector; and the alternative *Müllkonzept* based around the maximum promotion of recycling and waste reduction at source.[64]

THE 1991 PACKAGING ORDINANCE

A description of recycling and waste management in Hamburg would not be complete without mention of important recent developments within national waste legislation in Germany. The political pressure for higher rates of recycling and the difficulties faced by local government in recycling the diverse and growing component of waste derived from packaging, coupled with public opposition to new incineration plants, has led to radical new legislation at a national level in Germany. The new packaging ordinance was passed by the *Bundesrat* upper house in April 1991, and sets out stringent requirements for the collection and recycling of packaging by retailers and manufacturers. If the targets are met, 64 per cent of all glass, tin and aluminium packaging and 72 per cent of all paper, card, laminated packaging and plastics should be recovered by July 1995.[65] The law has instituted a 'dual waste system', where industry must organise the reclamation of reusable packaging waste and local authorities will continue to handle the collection and disposal of the remaining *Restmüll* from routine waste management and carry out their own recycling activities. The *Duale System Deutschland* (DSD) was set up by 400 companies in the private sector in order to comply with the new legislation. In the case of Hamburg the Wert GmbH has been set up as a subsidiary of the DSD which empties the yellow *Grüne Punkt* bins under contract to the city cleansing department though responsibility for site maintenance remains with the city of Hamburg. The yellow bins (or sacks) are used for plastics, laminated cartons, metal cans and foil, and the mixed recyclables are transported to centralized sorting plants before being distributed to different recycling industries. Yet the Hamburg Senate has been opposed to the introduction of the DSD system since they argue that there is no proven economic or ecological rationale and that the *Grüne Punkt* collections do not tackle the need for waste reduction at source.[66]

Products in the DSD *Grüne Punkt* system carry a green dot showing that their packaging can be recycled and qualifies for collection under the DSD scheme and firms within the scheme pay a licence fee passed onto the consumer through the higher cost of products. Under new legislation the German government is currently extending the DSD principle to consumer durables, particularly cars and white electrical goods, in order to force companies to take back and recycle as much as is technically feasible. It was originally estimated that an average of two pfennig would be placed on each item of packaging in order to finance seven million DM investment in 200 regional sorting facilities for the recyclable materials collected from retailers, yet the cost of DSD has steadily grown and may triple by 1995.[67] The DSD has guarantors for recycling different kinds of packaging; these guarantors organize the recycling and set up contracts with waste management firms, as illustrated in Figure 5.5. One factor behind the escalating cost of the system is that in order to meet the legally specified recycling quotas the DSD has to accept almost any contracts from recycling and waste management firms, however uncompetitive, and there has also been criticism of the efficiency of the organization itself in terms of the collection and sorting of materials.[68]

In the case of plastics, the guarantor is the Verwertungsgesellschaft für gebrauchte Kunstoffverpackungen (VGK). Plastics manufacturers such as Solvay, ICI, BASF and Bayer provide 37.5 per cent of the funding for the VGK. Plastics processors provide a further 37.5 per cent of the funding and the waste management industry provides the remaining 25.1 per cent. However, the VGK venture set up by the German packaging industry to recycle plastics waste has failed in the face of some 400,000 tonnes of collected materials exceeding the predicted total of 100,000 tonnes and VGK is now under investigation by the Frankfurt Public Prosecutors for the illegal export of plastic waste.[69] The export of German plastics abroad is now the focus of increasing concern, with *Grüne Punkt* plastics collected by the German public for recycling found on rubbish dumps as far afield as France and Indonesia. On 7 May 1993 activists from Greenpeace and FRAPNA (Rhône-Alpes Nature Protection Federation) delivered nearly three tonnes of German plastic waste, mostly *Grüne Punkt*, to the German Embassy in Paris in protest at the swamping of domestic markets and the impact of plastics disposal by landfill. The NIPSA recycling plant in the

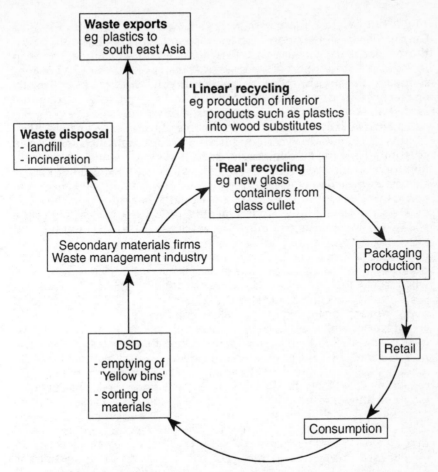

Figure 5.5 The Grüne Punkt system for packaging waste

Source: Bund für Umwelt und Naturschutz Deutschland, Berlin

Isère region of southern France receives up to 500 tonnes of German pack-aging materials a month, of which nearly 75 per cent are sent to landfill. In May 1993, the UK government lodged an official complaint with the European Commission since the UK plastics recycling sector was receiv-ing German material at zero cost and delivered free of charge, or in some cases with an incentive payment of up to £200 per tonne. As a result, UK imports of plastics grew by 450 per cent in 1992 alone, threatening British recycling efforts.[70] There is now growing conflict in the European Community over the degree of freedom for individual member states to enact environmental legislation that may undermine environmental poli-cy in other member states, since the disruption of recycling activity by the German DSD system effectively discourages the recycling and reuse of materials.[71]

Outside Western Europe, there is evidence of widespread exports of German waste to the poorer East European countries and to the Third World, particularly South East Asia. Indonesia, for example, has been inundated with foreign plastics wastes, mainly from the US, Western Europe and Japan. The influx of foreign wastes has had an adverse impact on the livelihoods of the 200,000 people who survive by collecting discarded materials from Indonesian garbage dumps for sale to local scrap merchants. In the case of Jakarta, the income of waste collectors has fallen from US $3 to just US $1.5 a day since 1992. Since April 1993 Indonesian port authorities have impounded more than 5000 tonnes of illegally imported plastic wastes and the Indonesian government's environment agency estimate that of the plastics imported for recycling, 30 per cent are non-recyclable and 10 per cent are actually hazardous wastes. The Indonesian Forum for the Environment (WALHI) are now calling for the plastic waste exporting countries to bear the costs of cleaning up and retrieving the wastes from Indonesia.[72]

A key problem is that the Grüne Punkt charge cannot adequately cover collection and recycling costs and that German industry has been overwhelmed by the amounts of waste collected by the public. The DSD and the German recycling industry has therefore been seeking the cheapest opportunities for the export of wastes and have paid foreign firms to accept wastes without ascertaining whether there are incentives or regulatory measures in their countries to ensure that wastes are actually recycled. There is concern that the European Commission has not adequately tackled the problem of waste exports for recycling: the new EC regulation 259/93 on the supervision of waste trade establishes in law a loophole allowing the export of all waste as long as it is claimed to be recyclable. The definition of recycling is broad enough to include incineration and also 'storage' of materials for eventual recycling and it is feared that all waste traders may now simply call themselves recyclers. A Greenpeace survey in 1992 revealed that 90 per cent of all waste trade schemes claimed some form of recycling or reuse for the exported toxic wastes.[73]

The German packaging ordinance has been fiercely criticized by German industry and economic interest groups within government who regard the new packaging controls as an unnecessary intrusion of the state and a threat to economic competitiveness. The UK based INCPEN representing 60 international companies from all sections of the packaging sector has lodged an official complaint with the European Commission, claiming that the law will distort trade between the EC member states and also threaten recycling industries in other countries. German industry is unable to cope with the collected materials, leading to dumping in other countries and a widespread collapse in recycling activities, as a result of a general depression of the secondary materials market.[74] The packaging lobby clearly fear the influence of the German legislation on a new packaging directive being drawn up by the EC and have been pushing for an extension of the definition of recycling to include energy recovery. The requirements that plastics and laminated packaging be recycled may force fillers, distributors and retailers to turn to other materials as one-way packaging, possibly resulting in a greater use of energy and resources. The German Coffee Association, if they were to use glass or tinplate instead of plastic film to pack their coffee, have already claimed that this would be the case.[75] This helps to explain the intense lobbying to change the definition of post-consumer recycling to include energy recovery and there is speculation that the German legislation will be altered to allow the incineration of their huge plastics stockpile.[76]

There are already signs that new legislation in France, Austria and the Netherlands is being modelled on the German system, and the first draft of the EC Packaging Directive in October 1991 called for a halt in the growth of the municipal waste stream to 150 kg per head and the recovery of 90 per cent of packaging waste within ten years.[77] The European Commission has taken an increasingly active role in waste management through a variety of directives aimed at the environmental impact of landfill and incineration and also the control of packaging waste. However, the Packaging Directive has been targeted by the UK government as an unnecessary intrusion by the European Commission into national policy making to be removed under the principle of subsidiarity.[78] These controversies within the European Community illustrate the complexity of resolving issues of national sovereignty with the evolving international legislative framework for environmental protection.

The legislation has also been attacked by the German environmentalist movement who claim that the underlying rationale of waste production at source has not been addressed, and that the volume of waste may continue to rise despite a larger proportion of packaging being recycled. The national environmental group BUND, for example, have urged consumers to boycott the system and have levelled a number of specific criticisms:[79]

- The *DSD system* is effectively a licence to produce waste and does not seriously address waste reduction at source.
- The cost of the *Grüne Punkt* licence is not related to the recyclability of different types of packaging but to volume. The levy on plastics should therefore be increased relative to other more recyclable packaging materials (though this may conflict with EC competition policy with respect to packaging materials).
- The system effectively endorses one-way packaging rather than encouraging the greater use of returnable packaging.
- There are still no adequate controls of packaging and waste materials containing toxic materials such as chlorine compounds and heavy metals.
- The costs of the *DSD system* are being borne by a captive consumer market raising the cost of living in a socially regressive manner.

The response of the German government and supporters of the DSD system to the mounting criticisms is focused on the belief that this is a transitional period during which the producers of waste and the recycling industry will need to adapt. The plastics industry in particular may be

seeking to 'ride out' the current wave of legislation without any fundamental changes in production patterns in the hope that the energy
recovery approach will prevail in the twenty first century.[80] It is argued
that the fundamental premise of the legislation, that the producers of
waste rather than the state, must be responsible for recycling, is still valid,
and that the DSD system has already resulted in a decline in non-recyclable and superfluous packaging.[81]

CONCLUSION

A comparison of London, Hamburg and New York shows that there are
systematic differences in policy under ostensibly similar developed
economies. The pattern of waste management in Hamburg and New York
suggests that an integrated form of local government, where the collection
and disposal of waste is carried out at a city wide level by one tier of government, is more conducive to recycling than a fragmentary
organizational structure where the collection and disposal of waste is carried out separately and is not accountable to one directly elected authority.
The wide difference in the extent of recycling in London and Hamburg
can be accounted for in part by the different administrative structures of
waste management and the degree of integration and institutionalization
of environmental policy into waste management, aided by the direct lobbying of the Hamburg Senate by the German Green Party elected under a
system of proportional representation. The environmentalist opposition
to incineration is also stronger in New York than in London, and this has
been significant in giving political impetus to the expansion of recycling.

The example of Hamburg also suggests that comprehensive recycling
is more expensive than waste disposal by landfill or incineration. In particular, the expense of extending recycling to the collection of putrescible
kitchen waste from homes without gardens shows that the promotion of
intensive recycling for high density urban areas raises political and economic dilemmas for public policy in the context of competing demands
for a range of essential services. The crisis in Hamburg's waste management has led to conflict between a focus on the need to control expenditure
on waste management but respond adequately to public demands for
recycling and an alternative environmentalist position seeking to maximize levels of recycling and waste reduction at source, coupled with a
clear rejection of the growing private sector role in waste management.
The conflict within the city is now set against a background of the trou-
bled efforts to control packaging at a national level through the *Grüne
Punkt* system and confusion over the economic and environmental impact
of the post-consumer recycling of packaging.

6

Conclusion

The pursuit of sustainable waste management as we enter the twenty first century must be placed within the context of diverse pressures on public policy from sources such as the revolution in information technology, the restructuring of the global economy, mass movements of people in search of a better quality of life, fluidity and uncertainty in political developments, and the difficulties facing the post-war Keynesian welfare state as an ageing population and intense fiscal pressures force a constant re-evaluation of the relative roles of the state and the market in modern capitalist societies.

In this book I have not examined in detail the technical merits of different recycling strategies, which are well covered elsewhere,[1] but I have sought to extend the waste management debate to the political and historical arena. It is important to recognize that there is no consensus over the most appropriate approach to waste management, reflecting the absence of any wider policy consensus over how best to tackle environmental problems since there are widely differing views over the underlying cause of the current crisis.[2] Decision making in the field of recycling and waste management cannot be reduced simply to a technical exercise since there are a range of subjective political judgements concerning the relative importance of different policy goals. These include employment generation, the protection of children's health and the control of public expenditure.

Since municipal waste accounts for no more than 4 per cent of the waste stream in developed economies, it is clear that the recycling of materials in household waste will not in itself stave off environmental catastrophe but there are a range of potential benefits, depending on which combination of recycling systems are adopted and the types of materials to be targeted:

- The environmental advantages of recycling and waste reduction in comparison with waste disposal by landfill or incineration.
- The generation of employment for low-skilled people in the context of persistent mass unemployment and the growing inequalities of opportunity within the labour market.

- The opportunity for public participation in recycling allows the raising of environmental awareness over the impact of individual behaviour on other communities threatened by profligate resource use, both now and in the future. Furthermore, participation may help people counter a sense of helplessness in the face of seemingly intractable global problems.

THE LIMITS TO COMPREHENSIVE POST-CONSUMER RECYCLING

There is a 'threshold' level of post-consumer materials recycling, beyond which it is extremely difficult to raise recycling rates without two main shifts in policy: first, the use of progressively more expensive forms of waste management such as kerbside collection schemes and the dual-bin system for the production of compost; and secondly, the greater use of legislative controls over production, retailing and marketing in order to strengthen the secondary materials market and tackle superfluous or toxic packaging and unrecyclable products. The maximum level of recycling achievable without a complete transformation of waste management policy at a national level is lower in inner urban areas because of a combination of difficulties:

- The lack of space both in the home and in common areas for the collection and storage of recyclable materials.
- The need to extend the collection of putrescible waste to households without gardens.
- The logistical aspects of waste collection involving the use of shared paladins.
- The high urban land values over which low return public service facilities such as recycling centres must compete with more profitable land uses.[3]

Many advocates of materials recycling have assumed that it is a means to lower the costs of waste management,[4] but most evidence suggests that high rates of recycling are more expensive than alternative means of waste disposal. This weakens the market-based argument that higher waste disposal costs as a result of tougher environmental standards will lead to a substantial increase in the recycling of waste, because recycling becomes progressively more expensive if it is extended to comprehensive kerbside recycling schemes collecting putrescible kitchen wastes, plastics and other materials. The key difference between recycling and incineration is that recycling is labour intensive whereas incineration is capital intensive. This presents a fundamental dilemma for developed economies with high labour costs and explains the shipping of mixed materials such as paper and plastics to Third World destinations for sorting. The expense of collecting and sorting materials from municipal waste in developed economies has risen over the post-war period, not only as a result of rising labour costs within the economy as a whole, but also as a result of the increasing complexity and non-recyclability of the waste stream, illustrated by the proliferation of different plastics polymers.

If the justification for recycling is moved beyond an economic rationale, as pursued by some London Boroughs reliant on the use of cheaper but less effective on-street collection facilities, to the provision of an environmental service through a comprehensive recycling strategy, as in Hamburg or New York, the key question is what the community is willing to pay in order to increase the level of materials recycling. This in turn suggests a complex trade-off between environmental protection and other aspects of public policy such as health, education and social services. Though there are theoretical arguments over the long-term economic benefits of recycling in comparison with waste disposal,[5] choices over different policy options have to be made *now* in a context of tight restrictions on public expenditure. The unresolved problem is how the higher costs of comprehensive recycling will be distributed in society and in particular, the identification of the polluter if the 'polluter pays principle' is to be followed. A key concern is that the use of direct charges on the supposed producers of waste through the increased cost of essential environmental services such as the collection and disposal of household waste is socially regressive, as shown in concern over the escalating flat rate *Müllgebühr* charge to individual households in Hamburg. If the expense of comprehensive materials recycling, including the recovery of environmentally significant plastics and putrescibles, is imposed under a simplistic 'polluter pays' framework, this may provoke a political backlash and alienate public concern over environmental issues:

> The question of who bears the cost, is a central concern in policy design. The impact on the consumer and the tax-payer is of great importance. Too large a burden on the consumer is likely to be regressive and carries the real risk of developing consumer resistance to environmental improvements... . Economists used to dismiss this objection with the argument that if the outcome is inequitable, governments can use their taxing powers to remedy this. The experience of the last decade must at least cause a pause for thought.[6]

THE NEED FOR WASTE REDUCTION

Waste reduction is clearly a more complicated and contentious aspect of waste management than the promotion of recycling from post-consumer waste. In the late 1970s the 'bring' system of recycling emerged based around the use of on-street collection facilities as a way to promote recycling by harnessing the environmental concerns of the public but enabling municipal authorities to escape the dilemma of escalating costs for public sector kerbside collection schemes. Yet the 'bring' system of recycling has raised new concerns over the local environmental impact of facilities such as bottle banks and the failure to make any substantial impact on the size of the municipal waste stream. Similarly, the reliance on kerbside recycling in the absence of waste reduction makes no serious long-term contribution to the reduction of pollution in the production process.[7]

However the consequences of the implementation of the waste reduction option based around greater government controls on industrial products and processes remains uncertain. Little is known about the long-term macro-economic effects, such as employment changes in different sectors of the economy and the possibility of reduced international competitiveness within the global economy. Waste prevention measures such as the maximum usage of returnable containers would also imply limitations on the liberalization of world trade to allow reduced transport costs within more self-sufficient regional economic systems. A serious attempt to reduce the size of the waste stream by government intervention in the production process to ensure that products are manufactured using clean technologies and are recyclable would imply a complete shift in industrial policy and an end to the uncontrolled consumerism of the post-war period.

Any attempt to actually reduce material output as part of a zero or negative growth waste reduction strategy would be politically and economically intolerable[8] and few policy makers or environmentalist groups seriously advocate this option. The waste reduction policy debate is therefore focused on how we can maintain or increase current levels of material prosperity through the effective utilization of clean technologies and other means to limit the usage of materials and energy in the production process. This is in essence what the sustainability debate is all about: the achievement of economic development and the elimination of poverty without environmental catastrophe in the twenty first century.

THE MARKET-LED INCINERATION PATH

The declining availability of landfill opportunities is leading to a re-emergence of incineration, with energy recovery as the most profitable short-term option for not only the private sector waste management industry but also for the packaging sectors most threatened by a policy emphasis on materials recycling and waste reduction at source.[9] It is the uncertain future of plastics which bests highlights the contemporary mismatch between the promotion of sustainable waste management at a global level and the unprecedented levels of prosperity achieved over the post-war period.

The recycling policy debate is now increasingly centred on the legitimate definition and focus of recycling within the hierarchy of recycling options from waste reduction at source to the recovery of materials and energy from post-consumer waste. It seems likely, for example, that the UK government will modify its 25 per cent recycling target to include energy recovery, with a scaled down specific target of 10 per cent for materials recovery as a realistic aim in the absence of substantial increases in public expenditure to cover the nation-wide adoption of comprehensive kerbside recovery schemes.[10] A powerful alliance is now emerging between the producers of waste (represented by the packaging lobby, the plastics industry and other sectors most threatened by recycling and waste reduction) and the disposers of waste (the private sector waste management industry and engineering companies looking for new markets following the collapse of the nuclear power station construction programme since the 1970s). This pro-incineration alliance is ranged against a diverse array of environmentalist groups and governments seeking to control the growth of the waste stream with an emphasis on materials recycling and waste reduction.

The market-based view of recycling policy assumes that low levels of recycling can be attributed simply to different forms of 'market failure'. Yet the policy emphasis and much of the literature has relied on a narrow focus on the cost externalities associated with waste disposal and have failed to recognize the greater profitability of incineration in comparison with comprehensive recycling programmes. A further concern is the contemporary market-based faith in 'green consumerism' as a substitute for community participation in environmental policy. The democratic link remains vital, to allow communities to choose which type of recycling and waste management strategy they wish to pursue, rather than that dictated by special interest groups lobbying the legislative process at a national and international level. In this respect, on the basis of the three case studies presented in this book, it is the waste management scenario facing London which is of greatest concern since it is difficult to conceive how an alternative strategy to the expansion of incineration could be articulated at a city-wide level. I wish to argue that environmental policy cannot be left to the vagaries and fluctuations of profit signals in the market place because a strategic approach in both space and time requires an active role for elected government in order to articulate the interests of the wider global community. A sustainable approach to waste management implies a rejection of the simplistic market dogma characteristic of the 1980s and a shift towards a public policy marked by a much more complex interrelationship between government and the private sector.

CONCLUSION

The notion of an inevitable shift to a post-industrial 'recycling society' of the future represents a fundamental *naïveté* on the part of many environmentalists and policy makers over the underlying constraints on recycling in a capitalist economy. Perhaps the most widely held environmentalist misconceptions are that 100 per cent recycling is possible and that comprehensive recycling is cheaper than waste disposal by other means. The current extent of recycling in London, Hamburg and New York lies well below technically achievable levels, but the cause of limited recycling is not attributable to lack of public participation, misguided policies pursued by municipal government, or 'market failure' in the costs of waste disposal. A distinction should be drawn between the local factors affecting the operational aspects of different recycling systems in practice, such as the extent of recycling infrastructure, and the role of underlying factors derived from the secondary materials market, the organizational and financial aspects of local government, and the degree of government control over the waste stream.

These underlying constraints form part of the wider context for recycling, which can only be altered significantly by changes in the legislative framework and the role of elected government in environmental policy. Yet this policy context for recycling is itself evolving in response to the combined impact of the political pressure to cut the costs of waste management as a corollary of the wider shift to market-based patterns of public policy and the need to respond to public demands for higher environmental standards. The result, somewhat paradoxically, is a shift in recycling policy away from the expense of comprehensive recycling of materials from post-consumer waste towards the increased use of incineration with energy recovery. This allows the development of profitable 'end of pipe' private sector environmental policy compatible with the current political and economic context, yet irrelevant to the need for more equitable and sustainable patterns of resource use at a global level.

Notes and References

INTRODUCTION

1   Hubbert, M K (1976) 'Outlook for fuel reserves' in Lapedes, D N (ed.), *Encyclopaedia of Energy* McGraw-Hill, New York
2   Young, J E (1991) *Discarding the Throwaway Society* Worldwatch Institute, Washington DC
3   *The Earth* (1992) publication by *The Guardian* in association with OXFAM to mark the UN Earth Summit held in Rio, June 1992
4   Melosi, M V (1981) *Garbage in the Cities: Refuse, Reform, and the Environment 1880–1980* Texas A&M University Press
5   Crowther, J (1974) 'Substitution – For communal or sectional benefit' paper presented to *The Conservation of Materials Conference*, 26–27 March, Harwell; Galbraith, J K (1958) *The Affluent Society* Penguin, Harmondsworth; Packard V (1960) *The Wastemakers* David McKay Company Inc , London
6   Agenda 21 quoted in Holmberg J; Thomson, K; and Timberlake, L (1993) *Facing the Future: Beyond the Earth Summit* Institute for Environment and Development/Earthscan, London, p 42
7   Brown, L R; Flavin, C; and Postel, S (1990) 'Picturing a Sustainable Society' in Brown L R (ed) *State of the World* Earthscan, London, pp 181–182; see also Kharbanda, O P and Stallworthy, E A (1990) *Waste Management: Towards a Sustainable Society* Greenwood Press, Westport, Connecticut
8   Brown, L R and Jacobson, J L (1987) *The Future of Urbanization: Facing the Ecological and Economic Constraints* Worldwatch Institute, Washington DC
9   Fincher (1989) 'The political economy of the local state' in Peet, R and Thrift, T (ed) *New Models in Geography* Volume One, pp 338–361; Sassen, S (1991) *The Global City: New York, London, Tokyo* Princeton University Press, Princeton, New Jersey
10 Young, J E (1991) *Discarding the Throwaway Society* Worldwatch Institute, Washington DC; Platt, B (1990) *Beyond 40%: Record-Setting Recycling and Composting Programs* Institute for Self Reliance/Island Press, Washington DC
11 A recent example is the address to the *20/20 vision* conference at the Goethe Institute London by the UK Minister of State at the Department of the Environment Timothy Yeo, 8 July 1993
12 The use of market-based policy instruments has emerged as a fundamental component in the new environmental consensus for sustainability (see 'A new emphasis on economic instruments' in *ENDS Report* 213, October 1992) and a large section of the environmentalist movement has also embraced the development of 'green consumerism' and market-based policies where the emphasis on individual responsibility and action has become easily incorporated into 'green consumerism' and the use of market-based policy instruments to modify consumer behaviour (see Chapter One in Gandy, M (1993) *Recycling and Waste: An exploration of contemporary environmental policy* Avebury, Aldershot)
13 See European Commission (1990) *Green Paper on the Urban Environment* European Commission Brussels; Elkin, T and McLaren, D (1991) *Reviving the City: towards sustainable urban development* Friends of the Earth and the Policy Studies Institute, London

THE MANAGEMENT OF MUNICIPAL WASTE

1   UK Department of the Environment (1992) *Digest of Environmental Statistics* HMSO, London
2   Wolf, N and Feldman, E (1991) *Plastics: America's Packaging Dilemma* Island Press, Washington DC
3   Stewart, J and Stoker, G (eds) (1989) *The Future of Local Government* Macmillan, London
4   Forester, W S (1991) 'Municipal Solid Waste Management in the United States' paper presented to the *London Waste Regulation Authority*, Annual Conference, 21 March; Gore, A (1992) *Earth in the Balance: Forging a new common purpose*, Earthscan, London; Klinski, S (1988) 'Besser als bisher – aber schlechter als nötig: Das Abfallgesetz 1986' in Institut für ökologisches Recycling *Abfall Vermeiden*, Fischer Taschenbuch Verlag, Frankfurt am Main, pp 123–132
5   Foley, G (1991) *Global Warming: Who is taking the heat?* Panos, London
6   Figures quoted in the *WARMER Bulletin*, August 1991. More alarmingly the National Society for Clean Air claim that UK landfill gas emissions are a greater contributor to the greenhouse effect than the transport sector ('waste regulators power restricted as landfill methane doubles' in *Environment Business* 7 April 1993)
7   Forester, W S (1991) 'Municipal Solid Waste Management in the United States' paper presented to the *London Waste Regulation Authority* Annual Conference, London, 21 March
8   Leipert, C and Simonis, U E (1990) *Environmental Damage – Environmental Expenditures: Statistical Evidence on the Federal Republic of Germany* Wissenschaftszentrum Berlin für Sozialforschung, Berlin
9   Forester W S (1991) 'Municipal Solid Waste Management in the United States' paper presented to the *London Waste Regulation Authority* Annual Conference, London, 21 March
10   'Landfill costs set to rise by up to 135%' *Environment Business*, 24 February 1993; Coopers and Lybrand (1993) *Landfill Pricing: Correcting Possible Market Distortions* HMSO, London
11   European Commission (1993) *Green Paper on Civil Liability* European Commission, Brussels
12   'Landfill directive resurrected and revised' *Environment Business* 14 July 1993; 'Waste operators rush to surrender licences' *ENDS Report* 220, 1993 pp 9–10; Friends of the Earth (1993) *Multi-million pound bill for taxpayers as waste companies race to dump pollution liabilities* Friends of the Earth, London
13   Richards, K M (1989) *Landfill Gas – Working with Gaia* Energy Technology Support Unit, Harwell. The UK has seen a steady expansion since the first operational system in 1980 with over 50 schemes running by early 1993 ('landmark for landfill gas utilisation' *ENDS Report* 221, June 93)
14   Tucker, D G (1977) 'Refuse destructors and their use for generating electricity: A Century of Development' *Industrial Archaeology Review*, 1, pp 74–96
15   Melosi, M V (1981) *Garbage in the Cities: Refuse, Reform and the Environment 1880–1980* Texas A&M University Press
16   Hoffman, R E (1986) 'Health Effects of Long-Term Exposure to 2378 Tetrachlorodibenzo-p-Dioxin' *The Journal of the American Medical Association*, April, pp 460–493; Spill, E and Wingert, E (eds) *Brennpunkt Müll* Sternbuch, Hamburg; Rappe, C; Choudary, G; and Keith, L H (eds) (1986) *Chlorinated*

Dioxins and Dibenzofurans in Perspective Lewis Publishers Inc, Chelsea, Michigan; The Women's Environmental Network (1989) *Dioxin: A Briefing* WEN, London

17 Newsday (1989) *Rush to Burn: Solving America's Garbage Crisis?* Island Press, Washington, DC

18 Pollock, C (1987) *Mining Urban Wastes: The Potential for Recycling* Worldwatch Paper 76, Worldwatch Institute, Washington DC

19 Underwood, J D; Hershkowitz, A; and de Kadt, M (1988) *Garbage: Practices Problems and Remedies* Inform Inc, US; Gore, A (1992) *Earth in the Balance: Forging a new common purpose* , Earthscan, London

20 Figures quoted in the *WARMER Bulletin,* January 1990

21 Royal Commission on Environmental Pollution (1993) *Incineration of Waste* HMSO, London

22 Tabasaran, O (1984) 'Emissions from the Incineration of Solid Waste' in Thome-Kozmiensky K J (ed) *Recycling International* EF Verlag für Umwelt und Technik, pp 83–88; Porteus, A (1990) 'Municipal Waste Incineration in the UK – What's Holding It Back' *Environmental Health,* pp 181–186

23 Hallbriter, G; Brautigen, K-R; Katzer, H; Braun, H; and Vogg, H (1984) 'Comparison of the Stack Emissions from Waste Incineration Facilities and Coal Fired Heating Power Stations' in Thome-Kozmiensky, K J (ed) *Recycling International* EF Verlag für Umwelt und Technik, Berlin, pp 58–63

24 Shiga, M (1975) 'Separate collection of household waste in Tokyo' paper presented to the *National Conference on Conversion of Refuse to Energy,* 3–5 November, Montreux

25 Gandy, M (1993) *Recycling and waste: An exploration of contemporary environmental policy* Avebury, Aldershot

26 Evans, R J (1991) *Death in Hamburg: Society and Politics in the Cholera Years 1830–1910* Penguin, Harmondsworth; Wohl, A S (1983) *Endangered Lives – Public Health in Victorian Britain* J M Dent & Sons Ltd, London

27 Melosi, M V (1981) *Garbage in the Cities: Refuse Reform and the Environment 1880–1980* Texas A&M University Press

28 Stewart, J and Stoker, G (eds) (1989) *The Future of Local Government* Macmillan, London

29 Melosi, M V (1981) op cit; Small, W E (1970) *Third Pollution: The National Problem of Solid Waste Disposal* Praeger, New York

30 Ambrose, P (1986) *Whatever happened to planning?* Methuen, London and New York; Lipietz, A (1992) *Towards a New Economic Order: Postfordism Ecology and Democracy* Polity Press, Cambridge; Martin, R (1986) 'Thatcherism and Britain's Industrial Landscape' in Martin, R and Rowthorn, B (eds) (1986) *The Geography of De-industrialisation* Macmillan, London, pp 238–291

31 Bacon, R and Eltis, W (1978) 'Too Few Producers' in Coates, D and Hillard, J (eds) (1986) *The Economic Decline of Modern Britain* Wheatsheaf Books Ltd, Brighton, pp 77–91; Stewart, J and Stoker, G (eds) (1989) *The Future of Local Government* Macmillan, London

32 Goddard, H C (1975) *Managing Solid Wastes: Economics Technology and Institutions* Praeger, New York; Savas, E (1977) 'Policy Analysis for Local Government: Public vs Private Refuse Collection?' *Policy Analysis* 3, pp 1–26; Savas, E S (1979) 'Public versus Private Refuse Collection: A Critical Review of the Evidence' *Urban Analysis* 6, pp 1–13

33 Melosi, M V (1981) op cit

34 Goddard, H C (1975) quoted in Gandy, M (1993) op cit p 40

35   Gandy, M (1993) op cit
36   'LA price cuts on refuse collection' *Environment Business*, 14 July 1993
37   Burck, C G (1980) 'There's Big Business in All That Garbage' *Fortune*,
      pp 106–112 Financial Times Survey (1990) *Waste Management*, 26 September
      1990
38   Financial Times Survey (1990) *Waste Management* p 1 quoted in Gandy, M
      (1993) op cit p 41

RECYCLING IN PERSPECTIVE

1   Barton, J R (1989) 'Recycling of packaging: source separation or centralised
      treatment' *Institute of Wastes Management* one day symposium, 4 October;
      Environmental Defense Fund (1987) *Coming Full Circle: Successful Recycling
      Today* EDF, New York; Hahn, E (1991) *Ecological Urban Restructuring*
      Wissenschaftszentrum Berlin für Sozialforschung; Platt, B (1990) *Beyond
      40%: Record-Setting Recycling and Composting Programs* Institute for Self
      Reliance/Island Press, Washington, DC;
2   Young, J E (1991) *Discarding the Throwaway Society* Worldwatch Institute,
      Washington, DC
3   Daly, H (1977) 'The Steady State Economy: What Why and How?' in
      Pirages D (ed) *The Sustainable Society: Implications for Limited Growth* Praeger,
      London and New York; Small, W E (1970) *Third Pollution: The National
      Problem of Solid Waste Disposal* Praeger, New York
4   Chapman, P F (1974) 'Energy costs of producing copper and aluminium
      from primary and secondary sources' paper presented to *The Conservation of
      Materials Conference*, 26–27 March, Harwell; Schertz, W (1984) 'Possibilities
      for separating and recycling disposable lined cardboard containers' in
      Thome-Kozmiensky, K J (ed) *Recycling International* EF Verlag für Umwelt
      und Technik Berlin, pp 784–789; City of New York Office of the Comptroller
      (1992) *Fire and Ice: How Garbage Incineration Contributes to Global Warming*
      submitted March 1992
5   Institut für ökologisches Recycling (1988) *Abfall Vermeiden* Fischer
      Taschenbuch Verlag, Frankfurt am Main; Institut für ökologisches Recycling
      (1989) *Ökologische Abfallwirtschaft: Umweltvorsage durch Abfallvermeidung* IföR,
      Berlin
6   Castle, K (1986) *The Recyclers' Guide to Greater London & Beyond: A Handbook
      for Resource Recovery* London Energy and Employment Network; Vogler, J
      (1981) *Work from Waste: Recycling wastes to create employment* Oxfam and
      Intermediate Technology Publications Ltd, London
7   Bundesamt für Umwelt Wald und Landschaft (1991) *Oekobilanz von
      Packstoffen* BUWAL, Bern; Stölting, P and Rubik, F (1992) *Übersicht über ökol-
      ogische Produktbilanzen* Institut für ökologische Wirtschaftsforschung (IÖW),
      Heidelberg
8   See for example Boustead, I and Hancock, G F (1984) 'Energy and recycling
      in glass and PET beverage container systems' Thome-Kozmiensky K J (ed)
      Recycling International EF – Verlag für Umwelt und Technik, Berlin,
      pp 751– 757
9   Virtanen, Y and Nilsson, S (1993) *Environmental Impacts of Waste Paper
      Recycling* Earthscan, London
10   Morris, J and Canzoni, D (1993) *Recycling Versus Incineration: An Energy
      Conservation Analysis* Sound Resource Management Group

11 Young, J E (1991) *Discarding the Throwaway Society* Worldwatch Institute, Washington, DC, p 26

12 Chandler, W U (1983) *Materials Recycling: The Virtue of Necessity* Worldwatch Paper 56, Worldwatch Institute, Washington, DC; Dyson, B H (1974) 'Efficient utilisation of materials – One answer to our balance of payments problem' paper presented to *The Conservation of Materials Conference*, 26–27 March, Harwell

13 Hayes, D (1978) *Repairs Reuse Recycling – First Steps Toward a Sustainable Society* Worldwatch Paper 23, Worldwatch Institute, Washington, DC; Risch RWK (1978) 'The Raw Material Supply of the European Community: The Importance of Secondary Raw Materials' *Resource Policy*, pp 181–188

14 Holmberg, J; Thomson, K; and Timberlake, L (1993) *Facing the Future: Beyond the Earth Summit* International Institute for Environment and Development/Earthscan, London

15 Sweetman, J (1979) 'Making Paper by Hand' *Appropriate Technology* Intermediate Technology Publications Ltd, UK; Vogler J (1980) 'Glassmaking as in the Time of Yore' *Materials Reclamation Weekly*, 18 October, pp 24–26

16 Castle, K (1986) *The Recyclers' Guide to Greater, London ...& Beyond: A Handbook for Resource Recovery* London Energy and Employment Network

17 Ossenbrügge, J (1988) 'Regional Restructuring and the Ecological Welfare State – Spatial Impacts of Environmental Protection in West Germany' *Geographische Zeitschrift 76*, pp 78–96

18 Bahro, R (1986) *Building the Green Movement* Gay Men Publishers, London

19 Lovins, A (1977) *Soft Energy Paths* Penguin, Harmondsworth; Young, J E (1991) *Discarding the Throwaway Society* Worldwatch Institute, Washington, DC

20 Norton, R (1986) *Community Scale Recycling* unpublished M Phil Thesis, University of East Anglia

21 See Chapter One in Gandy, M (1993) op cit

22 Kopytziok, N and Oswald, R (1988) 'Eine Abfall-Odysse' in Institut für ökologisches Recycling, *Abfall Vermeiden* Fischer Taschenbuch Verlag, Frankfurt am Main, pp 23–28; Moser, M (1990) *Environmental relief through waste prevention* unpublished paper, Institut für ökologishes Recycling, Berlin

23 See Tattersley, P (1990) 'Recycling in Denmark' *Environmental Health*, June 1990, pp 158–160

24 See for example Bunyard, P and Morgan-Grenville F (1987) *The Green Alternative: Guide to Good Living* Methuen, London

25 Waste Management Advisory Council (1976) *Report on Waste Paper Collection by Local Authorities*, HMSO, London

26 Gandy, M (1991) 'Environmental policy and local government' in McCoshan, A and Pinto, R R (eds) (1991) *Local Services: Past Experiences and Possible Scenarios*, pp 46–52 Geography Discussion Papers, New Series No 24, Graduate School of Geography, The London School of Economics

27 Hartmann, M and Rudolphi, M (1989) 'Zur Frage der Akzeptanz unterschiedlicher Entsorgungssyteme im Raum München' in Haas H-D (ed) (1990) *Müll – Untersuchungen zur Problemen der Entsorgung und des Rohstoffrecycling* Münchner Studien zur Sozial-und Wirtschaftsgeographie, Band 35, Verlag Michael Lassleben Kallmunz, Munich

28 Tattersley, P (1990) 'Recycling in Denmark' *Environmental Health*, June 1990, pp 158–160

29 Birley, D (1993) 'Does the Blue Box have a future?' *WARMER Bulletin 36*, pp 10–11

30   Brown, R; Coggins, C; and Cooper, D (1990) 'Salvaging the Waste' *The
     Surveyor*, 17 May, pp 20–22; UK Gallup Poll, January 1990; UK ICM Poll,
     January 1990
31   UK Warren Spring Laboratory (1993) *Overview of the Impact of Source
     Segregation Schemes on the Household Waste Stream in the UK and their
     Relevance to the Government's Recycling Target*, WSL
32   Gandy, M (1993) op cit
33   Interview with the Assistant Manager for the West London Waste
     Authority, Ray Brown, 16 February 1990; Site visit to the East Harlem mate-
     rials recovery facility with the Project Manager for the Contracts Unit of the
     City of New York Department of Sanitation, Penny Miller, 3 September 1993
34   The Environmental Defense Fund (1987) *Coming Full Circle: Successful
     Recycling Today* EDF, New York; Goldman, E A and Izeman, M A (1990) *The
     New York Environment Book* Island Press, Washington, DC; Uitgeverij
     Milieudefensie (1991) *Verpakkingen* Verenigung Milieudefensie, Amsterdam
35   'Time to shrink wrap' *The Guardian*, 9 October 1992
36   Jessop, B (1991) *Fordism and post-fordism: a critical reformulation* Lancaster
     Regionalism Group, Working Paper 41, University of Lancaster; Lipietz, A
     (1992) *Towards a New Economic Order: Postfordism Ecology and Democracy*
     Polity Press, Cambridge
37   Melosi MV (1981) op cit p 190
38   Schmidt-Alck, S and Strenge, U (1988) 'Glas – ein Plädoyer für Mehrweg' in
     Institut für ökologisches Recycling, *Abfall Vermeiden* Fischer Taschenbuch
     Verlag, Frankfurt am Main, pp 45–53; Small, W E (1970) *Third Pollution: The
     National Problem of Solid Waste Disposal* Praeger, New York
39   Gehrke, C (1988) 'Möglichkeiten und Grenzen des abfallarmen Einkaufs' in
     Institut für ökologisches Recycling, *Abfall Vermeiden* Fischer Taschenbuch
     Verlag, Frankfurt am Main, pp 99–106
40   Friends of the Earth (1993) *Overpackaging: Wasting Money Wasting Resources*
     Friends of the Earth, London
41   Jordan, G and Wessel, K (1988) 'Papier – Baume BILD und Altpapier' in
     Institut für ökologisches Recycling, *Abfall Vermeiden* Fischer Taschenbuch
     Verlag, Frankfurt am Main, pp 29–37; Kopytziok, N (1988a) 'Grundlagen für
     eine umfassende Betrachtung der Abfallproblematik' in Institut für ökolo-
     gisches Recycling, *Abfall Vermeiden* Fischer Taschenbuch Verlag, Frankfurt
     am Main, pp 17–23
42   Wirth, B (1988) 'Kunstoffe: Problemaufriss bei der Herstellung der
     Verwendung und Beseitigung' in Institut für ökologisches Recycling, *Abfall
     Vermeiden* Fischer Taschenbuch Verlag, Frankfurt am Main, pp 38–44;
     Bogenschütz, P (1991) 'Ökologie und Design – die Rolle des Designs bei der
     Entwicklung umweltverträglicher Produkte' in Institut für ökologisches
     Recycling *Perspektive Abfallvermeidung*, IfÖR, Berlin
43   Research by Franklin Associates cited in Blumberg, L and Gottlieb, R (1989)
     War on Waste: Can America Win Its Battle With Garbage Island Press,
     Washington, DC
44   Wolf, N and Feldman, E (1991) *Plastics: America's Packaging Dilemma* Island
     Press, Washington, DC; Pollock, C (1987) *Mining Urban Wastes: The Potential
     for Recycling* Worldwatch Paper 76, Worldwatch Institute, Washington, DC
45   Interview with Ute Dibbert, Bund für Umwelt und Naturschutz eV, Berlin
     13 September 1993
46   Wirth B (1988) 'Kunstoffe: Problemaufriss bei der Herstellung der
     Verwendung und Besietigung' in Institut für ökologisches Recycling, *Abfall
     Vermeiden* Fischer Taschenbuch Verlag, Frankfurt am Main, pp 38–44

47 Blumberg, L and Gottlieb, R (1989) op cit
48 Die Grünen (1989) *Verzicht auf PVC und Chlorchemie* Die Grünen Bonn; Wirth, B (1988) 'Kunstoffe: Problemaufriss bei der Herstellung der Verwendung und Besietigung' in Institut für ökologisches Recycling, *Abfall Vermeiden* Fischer Taschenbuch Verlag, Frankfurt am Main, pp 38–44
49 Greenpeace International Toxic Trade Campaign (1993) *We've been had! Montreal's Plastics dumped overseas* Greenpeace International, Montreal, Quebec
50 'Packaging data "null and void"' *Environment Business*, 3 June 1993
51 Greenpeace International Toxic Trade Campaign (1993) *We've been had! Montreal's Plastics dumped overseas* Greenpeace International, Montreal, Quebec, p 7
52 'European environmentalists agree packaging agenda' *ENDS Report* 215, December 1992
53 Heering, H; van Geldner, J W; and Blaauw, J A (1992) *Tetra Pak in Eastern Europe* Friends of the Earth, Netherlands, Amsterdam
54 Schmidt-Alck, S and Strenge, U (1988) 'Glas – ein Plädoyer für Mehrweg' in Institut für ökologisches Recycling, *Abfall Vermeiden* Fischer Taschenbuch Verlag, Frankfurt am Main, pp 45–53. See also Friends of the Earth, (1993) *Bring Back the 'Bring Back': The environmental benefits of reusable packaging* FoE, London
55 Bojkow, E (1984) 'Possibilities and Limitations of Waste Reduction in Beverage Packaging' in Thome-Kozmiensky, K J (ed) *Recycling International*, EF – Verlag für Umwelt und Technik, Berlin, pp 725–732
56 Golding, A (1989) 'Energieverbrauch und dessen Berechnung in vergleichenden Ökobilanzen von Getränkeverpackungen' paper presented to the conference, *Ökologische Abfallwirtschaft*, 30 November to 2 December, Technische Universität, Berlin
57 Burkard, T; Jordan, G; and Schwensen, J (1988) 'Ökologisches Abfallkonzept für West-Berlin: Im Vergleich zu den Auswirkungen einer geplanten Müllverbrennungsanlage' in Institut für ökologisches Recycling, *Abfall Vermeiden* Fischer Taschenbuch Verlag, Frankfurt am Main, pp 145–153
58 British Plastics Federation (1979) *Technical Factors Governing the Recycling of Plastics* British Plastics Federation, London; Flood, M (1993) 'Over-packaging or over–consumption?' in the *WARMER Bulletin* 37, pp 16–17; The Industry Council for Packaging and the Environment (1987) *Packaging Saves Waste* INCPEN, London
59 Falk, H (1988) 'Weissblechverpackungen: Umwelt belastungen bei der Produktion und Entsorgung' in Institut für ökologisches Recycling, *Abfall Vermeiden* Fischer Taschenbuch Verlag, Frankfurt am Main, pp 54–63
60 Hooper, G (1984) 'Possibilities of state control – development tendencies' in Thome-Kozmiensky, K J (ed) *Recycling International* EF Verlag für Umwelt and Technik, Berlin, pp 699–701; Troge, A (1984) 'Effects of a charge on drinks packaging on the environment and on business' in Thome-Kozmiensky (ed) *Recycling International* E F - Verlag für Umwelt und Technik, Berlin
61 Figures quoted in the *WARMER Bulletin*, May 1993
62 Troge, A (1984) 'Effects of a charge on drinks packaging on the environment and on business' in Thome-Kozmiensky (ed) *Recycling International*, E F Verlag für Umwelt und Technik, Berlin
63 Research by the German Society for Research into the Packaging Market quoted in the *WARMER Bulletin*, August 1990

64　Blair, I C (1987) 'Pulling the weight of recycling' in *Beverage World*, June, pp 28–34; Bremme, H C (1984) 'Economic effects of planned governmental packaging regulations on a food chain' in Thome-Kozmiensky, K J (ed) *Recycling International*, EF - Verlag für Umwelt und Technik, Berlin, pp 713–718

65　Report of the Moreland Act Commission on the Returnable Container Act submitted in March 1990 to Governor Mario Cuomo, New York State

66　Gandy, M (1993) op cit

67　See Hoy, S M and Robinson, M C (1979) *Recovering the Past: A Handbook of Community Recycling Programmes, 1890–1945* Public Works Historical Society Chicago; Temple, F C (1943) 'Wealth from waste' *The Journal of the Institution of Municipal and County Engineers*, 69, pp 405–13

68　Interview with the Assistant Cleansing Manager for the London Borough of Enfield, Peter Joyce, 12 June 1990

69　Interview with Dr Eckhard Willing, Umwelt Bundesamt, Berlin, 16 September 1993

70　See for example Taira, K (1969) 'Urban Poverty Ragpickers and the "Ants' Villa" in Tokyo' *Economic Development and Cultural Change*, 17, pp 153–177

71　See Haas, H -D and Sagawe, T (1989) 'Kommunale und informelle Abfallwirtschaft in Santo Domingo/Dominikanische Republic' in Haas, H - D (ed) (1990) *Müll – Untersuchungen zur Problemen der Entsorgung und des Rohstoffrecycling*, Münchner Studien zur Sozial-und Wirtschaftsgeographie, Band 35, Verlag Michael Lassleben Kallmunz, Munich

72　De Young, R (1986) 'Some Psychological Aspects of Recycling: The Structure of Conservation Satisfactions' *Environment and Behaviour*, 18, pp 435–449

73　The Environmental Defense Fund (1987) *Coming Full Circle: Successful Recycling Today* EDF, New York

74　Vining, J and Ebreo, A (1990) 'What Makes A Recycler? A Comparison of Recyclers and Nonrecyclers' *Environment and Behaviour*, 22 , pp 55–73

75　Ingenieurgemeinschaft Technischer Umweltschutz (1989) *Verfahren zur Reduktion des Hausmüllaufkommens* Ingenieurgemeinschaft Technischer Umweltschutz, Berlin; Pieters, R G M and Verhallen, T M M (1986) 'Participation in Source Separation Projects: Design characteristics and per-cieved costs and benefits' *Resources and Conservation*, 12, pp 95–111; Scheffold, K (1984) 'Source separation of packaging – materials and recovery possibilities' in Thome-Kozmiensky, K J (ed) *Recycling International*, EF Verlag für Umwelt und Technik, Berlin, pp 796–801

76　Scheffold, K (1984) 'Source separation of packaging – materials and recovery possibilities' in Thome-Kozmiensky K J (ed) *Recycling International*, EF-Verlag für Umwelt und Technik, Berlin, pp 796–801

77　Cooper, J (1989) 'Recycling' paper presented to the SEEDS (South East Economic Development Strategy) conference *Green Plan for the South East*, Brighton, 19–20 October; Forester, W S (1991) 'Municipal Solid Waste Management in the United States' paper presented to the *London Waste Regulation Authority*, Annual Conference, London, 21 March. *The Guardian*, 23 October 1991; Young. J E (1991) *Discarding the Throwaway Society*, Worldwatch Institute, Washington, DC

78　Gandy, M (1993) op cit

79　Darnay, A and Franklin, W E (1972) *Salvage Markets for Materials in Solid Wastes* US Environmental Protection Agency, Washington, DC

80　Melosi, M V (1981) op cit

81　Newell, J (1990) 'Recycling Britain' *New Scientist*, 8 September, pp 46–49

82 Department of the Environment (1991) *Waste Management Paper No 28: Recycling*
83 'Waste tax plan offers lifeline to recyclers' *The Observer*, 16 May
84 Though there is uncertainty over the extent of landfilling of paper collected for recycling, the practice has been alleged by Friends of the Earth and a number of UK local government officers with responsibility for recycling
85 Jordan, G and Wessel, K (1988) 'Papier – Baume BILD und Altpapier' in Institut für ökologisches Recycling, *Abfall Vermeiden* Fischer Taschenbuch Verlag, Frankfurt am Main, pp 29–37
86 Figures quoted in the *Paper Market Digest*, June 1993, p 4
87 Gandy, M (1993) op cit
88 Ball, R and Matthews, R (1988) 'Glass recycling by local authorities: an economic evaluation' *Resources policy*, 14, pp 205–217
89 See, for example, Warren Spring Laboratory (1993) *Overview of the Impact of Source Segregation Schemes on the Household Waste Stream in the UK and their Relevance to the Government's Recycling Target* WSL, UK
90 Interview with the Senior Planner for the London Borough of Sutton, Graham Dean, 10 January 1990
91 Interview with the Assistant Recycling Officer for the London Borough of Richmond, Sally Lawes, 19 June 1990
92 See Gandy, M (1993) 'Market-based policy instruments for environmental protection' *Environmental Information Bulletin*, August 1993; Nicolaisen, J; Dean, A and Hoeller, P (1991) *Economics and the environment: a survey of issues and policy options* OECD Economic Studies no 16, OECD, Paris
93 Owens, S; Anderson, V; and Brunskill, I (1990) *Green Taxes* Green Paper no 2, The Institute for Public Policy Research, London.Quoted in Gandy (1993) op cit, p 25
94 Opschoor, J B and Vos, H B (1989) *Economic Instruments for Environmental Protection* Organisation for Economic Co-operation and Development, Paris
95 Turner, R K (1992) *An Economic Incentive Approach to Regulating the Throwaway Society* CSERGE Working Paper PA 92-05, UEA and UCL, p 4
96 Young, J E (1991) *Discarding the Throwaway Society* Worldwatch Institute, Washington, DC, p 30 emphasis added. See also Curlee, T R (1986) *The economic feasibility of recycling: A case study of plastic wastes* Praeger Publishing, New York; Pollock, C (1987) *Mining Urban Wastes: The Potential for Recycling* Worldwatch Paper 76, Worldwatch Institute, Washington DC; Williamson, J B P (1974) 'An industrialist's view of the recovery and recycling of materials' paper presented to *The Conservation of Materials Conference* 26–27 March, Harwell
97 See, for example, Moody, R (1991) *Plunder* published by People Against RTZ and its Subsidiaries in London and the Campaign Against Foreign Control of Aotearoa, New Zealand; Monroe, R and Moody, R (1993) 'Mining's New World Order Neo-colonialism by any other name' *Higher Values*, Newsletter of Minewatch, January 1993, pp 11–13
98 'Little doubt about landfill levy' *Environment Business*, 21 April 1993
99 Pollock, C (1987) *Mining Urban Wastes: The Potential for Recycling* Worldwatch Paper 76, Worldwatch Institute, Washington DC, p 27. See also UK Department of the Environment (1990) *This Common Inheritance* HMSO, London; UK Department of Trade and Industry/Department of the Environment (1992) *Economic Instruments and Recovery of Resources from Waste* HMSO, London; Turner, R K (1990) *Towards an Integrated Waste Management Strategy* Key Environmental Issues: Number Eleven in a Series, British Gas, London; Pearce, D and Turner, R K (1992) *The Economics of*

Packaging Waste Management: Conceptual Overview CSERGE Working Paper WM 92-03, University of East Anglia

100 'Disputed New Role for Polls: Putting a Price Tag on Nature' *New York Times,* 6 September 1993

101 See Chapter One in Gandy, M (1993) op cit

102 Note for example how multinational corporations are currently hardening their views on property rights to control the results of plant and animal breeding in order to secure long–term profitability within the global agricultural sector in the face of economic uncertainty. See Holmberg, J; Thomson, K; and Timberlake, L (1993) *Facing the Future: Beyond the Earth Summit* International Institute for Environment and Development/Earthscan, London

LONDON

1   London Waste Regulation Authority (1993) *A Prospect for Change: Interim Waste Management Statement for Greater London 1993* draft submitted in April 1993

2   UK Department of the Environment (1990) *This Common Inheritance* HMSO, London

3   Quoted in the Herbert Commission (1960) *Report of the Royal Commission on Local Government in Greater London* HMSO, London, p 168

4   Metropolitan Commission of Sewers (1849) *A report to the survey committee on street cleansing* Reynell and Weight, London, p 4 quoted in Gandy (1993) *Recycling and Waste: an exploration of contemporary environmental policy* Avebury, Aldershot

5   Barker, B (1946) *Labour in London: A study in municipal achievement* George Routledge, London

6   Winter, J (1989) 'The "Agitator of the Metropolis": Charles Cochrane and Early-Victorian Street Reform' *The London Journal,* 14, pp 29–42

7   Goodrich, W (1904) *Refuse disposal and power production* Archibald Constable and Company Ltd, London

8   Local Government Board (1915) *Return as to scavenging in Urban Districts 1914,* HMSO, London, p xi

9   Metropolitan Borough of Shoreditch (1925) *Report to the Electricity Committee on the Disposal of Refuse*

10 Ministry of Health (1929) *Report of an investigation into the Public Cleansing service in the administrative county of London* HMSO, London

11 Herbert Commission (1960) *Report of the Royal Commission on Local Government in Greater London* HMSO, London

12 Public Health Engineering was to be one of seventeen departments in the GLC, and in addition to refuse disposal it was also responsible for main drainage, sewage purification and trade effluent control, along with land drainage, river pollution control and Thames flood prevention. However the responsibility for water was to be taken away in 1973–4 with the creation of the unelected Thames Water Authority, and this marked an important early reversal in the strategic GLC role for public service provision in the capital. See Mukhopadhyay, A K (1975) 'The Politics of London Water' *The London Journal,* 1, pp 207–226

13 Townend, W K (1982) 'Waste Disposal in Greater London – Operational Developments since 1965' *Wastes Management,* 72, pp 229–235

14 Vick, E H and Flintoff, F L D (1966) 'Refuse Disposal in Greater London' paper presented to *The Institute of Public Cleansing Annual Conference*, 7–10 June, Bournemouth

15 Greater London Council (1980) *Talking Rubbish: The GLC's new strategy for dealing with London's waste*

16 Cooper, J (1981) *Rubbish: A review of the GLC's solid waste disposal plans 1965–1995* Occasional Paper no 9, School of Geography, Kingston Polytechnic

17 Cooper, J (1981) op cit

18 Greater London Council (1980) *Talking Rubbish: The GLC's new strategy for dealing with London's waste*

19 Coggins, P C; Cooper, A D; and Brown, R W (1989) 'Civic Amenity Waste Disposal Sites: The Cinderella of the Waste Disposal System' paper presented to *The Institute of British Geographers*, Annual Conference, Coventry Polytechnic 5 January

20 Townend, W K (1982) 'Waste Disposal in Greater London – Operational Developments since 1965' *Wastes Management*, 72, pp 229–235

21 UK Department of the Environment (1971) *Refuse Disposal* HMSO, London

22 Greater London Council (1984) *Recycling for Employment Conference* Conference held at County Hall, 10 November

23 Greater London Council (1984) *Recycling – Support for recycling workshops* report of the Public Services and Fire Brigades Committee, 14 March

24 Greater London Council (1983) *Landfill sites: methane generation and recovery* Greater London Council (1983) *No Time to Waste: A limited Planning Statement for Waste Regulation Recovery and Disposal 1983–2003*

25 Greater London Council (1984) *Evidence to the Royal Commission on Environmental Pollution* Report of the Public Services and Fire Brigade Committee Greater London Council (1985) *European Community Environmental Policy and British Local Government* Report of the Environmental Panel, 13 September

26 Turner, R K and Thomas, C (1982) 'Source separation recycling schemes' *Resources Policy*, March, pp 13–24

27 Townend, W K (1982) 'Waste Disposal in Greater London – Operational Developments since 1965' *Wastes Management*, 72, pp 229–235

28 Greater London Council (1985) *Recycling – The Council's response to the Trade and Industry Committee's Report – "The Wealth of Waste"* report of the Environmental Panel, 1 February 1985

29 UK Department of the Environment (1983) *Streamlining the Cities* HMSO, London, quoted in Gandy (1993) op cit, p 105

30 Greater London Council (1984) *Preparation of the Council's case against main abolition legislation for waste disposal in London – parliamentary briefing note* confidential paper for weekly campaign meeting on 20 August by Controller of Operational Services

31 Greater London Council (1984) *Streamlining the Cities: Response by the GLC Conservative Group* quoted in Gandy (1993) op cit, p 108

32 Interview with the Project Officer (Cleansing) for the City of Westminster, Caesar Voûte, 15 May 1990

33 Interview with the Waste Planning Officer for the East London Waste Authority, Tom Butterfield, 13 June 1990

34 Interview with the Recycling Development Officer for the London Borough of Camden, Mike Newport, 15 June 1990

35   Interview with the Highways and Cleansing Manager for the London Borough of Barking and Dagenham, Terry Mirams, 7 June 1990. Interview with the Technical Services Officer for the London Borough of Hammersmith and Fulham, Tony Talman, 7 June 1990

36   Interview with the Recycling Development Officer for the London Borough of Camden, Mike Newport, 15th June 1990. Interview with the Senior Planner for the London Borough of Sutton, Graham Dean, 10 January 1990

37   Gandy (1993) op cit

38   Interview with the Cleansing Manager for the London Borough of Islington, Bob Lapsley, 18 April 1990. Interview with the Cleansing Officer for the London Borough of Kensington and Chelsea, Moira Billinge, 16 May 1990. Interviews with the Environmental Development Manager, Peter Brooker, and the Cleansing Officer, David Masters, for Globe Town Neighbourhood, London Borough of Tower Hamlets 20 June 1990 Interview with the Project Officer (cleansing) for the City of Westminster, Caesar Voûte, 15 May 1990

39   Greater London Council (1985) *Recycling – Composting at source* Report of the Public Services and Fire Brigades Committee, 7 May

40   Cooper, J (1990) 'Recycling Incentive Payments' unpublished draft paper presented to the Local Authority Recycling Advisory Committee in October 1990

41   Local Authorities Management Services and Computer Committee (1971) *Waste Paper Salvage* LAMSAC, London Waste Management Advisory Council (1976) *Report on Waste Paper Collection by Local Authorities* HMSO, London

42   Meacher, M (1975) 'Riddle of the great waste paper chase' *The Guardian*, 3 April

43   John, B (1974) 'Recycling politics' *New Scientist*, 10 January, p 60

44   UK Department of the Environment/Department of Trade and Industry (1974) *War on Waste: A Policy for Reclamation* HMSO, London

45   Townsend, E (1975) 'Ironies of the waste paper mountain' *The Times*, 16 July 1975

46   *Municipal Engineering*, 13 June 1975

47   Greater London Council (1975) *Solid Wastes Management – Waste Paper* Report of the Public Services Committee, 10 June

48   *Brentford and Chiswick Times*, 12 June 1975

49   *Materials Reclamation Weekly*, 28 August 1976

50   *Barking and Dagenham Post*, 14 April 1976

51   Letter from the Controller of Operational Services to the GLC Public Health Engineering Department, 22 May 1974

52   'Controversy follows grant for newsprint recycling mill' *ENDS Report* 217, February 1993

53   Interview with the Waste Disposal Manager for the London Borough of Havering, Ralph Johnson, 13 June 1990

54   *Materials Reclamation Weekly*, 13 January 1990

55   Gandy, M (1993) op cit

56   Greater London Council (1984) *'The Wealth of Waste': Evidence to the House of Commons Trade and Industry Committee Inquiry into Waste Recycling* report of the Public Services and Fire Brigade Committee, 19 June

57   Greater London Council (1985) *Recycling – A review of the recycling of beverage containers and proposals for further re-use and recycling* Report of the Public Services and Fire Brigade Committee, 31 January

58   Edwards (1974) opening address by the Chair of the GLC Public Services Committee to the *Symposium on Solid Waste Disposal*, 29 October 1974, p 67 quoted in Gandy (1993) op cit

59   Townsend, E (1974) 'Recycling glass: a not so simple task' *The Times*, 4 September

60   London Boroughs Association (1982) *Bottle Banks: Expansion of the Bottle Bank Scheme in London and the Surrounding Area*

61   Townsend, P; Corrigan, P; and Kowarzik, U (1987) *Poverty and Labour in London* Low Pay Unit, London

62   Interview with the Environment Promotions Officer for the London Borough of Bromley, Simon Bussel, 18 May 1990

63   Interview with the Recycling Development Officer for the London Borough of Camden, Mike Newport, 15 June 1990. Interview with the Project Officer (cleansing) for the City of Westminster, Caesar Voûte, 15 May 1990

64   See *What's On*, 17 January 1990; *Time Out*, 1 January 1991; *Evening Standard*, 5 July 1993 and many other examples

65   See, for example, Breckon, B (ed) (1990) *The Green Guide to London* Simon and Schuster, London and Sydney

66   Friends of the Earth (1989) *Recycling projects and the employment training scheme* UK 2000/Friends of the Earth, London

67   'Protests rise over German waste exports' *Environment Business*, 19 May 1993

68   The analysis of recycling in London presented in Tables 3.2 3.3 and 3.4 and in Figures 3.5 and 3.6 is based on a survey of recycling in the, London Boroughs I carried out in 1990 funded by the Economic and Social Research Council. For fuller details see Gandy, M (1993) *Recycling and waste: an exploration of contemporary environmental policy* Avebury, Aldershot

69   UK Department of the Environment (1991) *This Common Inheritance: The First Year Report* HMSO, London

70   Gandy (1993) op cit

71   Greater London Council (1975) *Solid Wastes Management – Waste Paper* Report of the Public Services Committee, 10 June

72   Gandy (1993) op cit

73   Friends of the Earth (1991) *A Survey of Local Authority Recycling Schemes in England and Wales* Friends of the Earth, London

74   'Key report on economic instruments and recycling' *ENDS Report* 215, December 1992

75   Telephone Interview with the Recycling Officer for the London Borough of Kensington and Chelsea, Sharon Deane, 17 August 1993

76   Gandy (1993) op cit

77   Audit Commission (1984) *Securing Further Improvements in Refuse Collection* HMSO, London

78   Coopers and Lybrand Associates (1981) *Service Provision and Pricing in Local Government. Studies in Local Environmental Services* HMSO, London. Savas, E S (1979) 'Public versus Private Refuse Collection: A Critical Review of the Evidence' *Urban Analysis*, 6, pp 1–13

79   Coopers and Lybrand Associates (1981) *Service Provision and Pricing in Local Government Studies in Local Environmental Services* HMSO, London. Forsyth, M (1980) *Re-Servicing Britain* Adam Smith Institute, London. Green, D G (1987) *The New Right: The counter revolution in political economic and social thought* Wheatsheaf Books, Brighton

80   Evans, R (1991) 'French ready to pick up pieces' *The Financial Times*, 26 June

81   The Economist (1989) *Water industry: Storming the barricade*, 14 October, pp 45–46
82   Coopers and Lybrand (1981) op cit
83   Gandy, M (1993) op cit
84   UK Department of the Environment (1989) *The Role and Function of Waste Disposal Authorities* HMSO, London
85   Personal communication with Steve Clark of the Land Wastes Division of the Department of the Environment, 1992
86   Lambert, E and Laurence, D (1990) 'Environmental Protection Bill: All Change for Waste Management' *Wastes Management*, March, pp 187–200
87   Wakeford, T (1990) 'Britain's Green Bill' *WARMER Bulletin*, 26, p 13
88   Interview with the Assistant Cleansing Manager for the London Borough of Enfield, Peter Joyce, 12 June 1990. Interview with the Principal Engineer (Services) in the London Borough of Waltham Forest, Barry Higgs, 22 May 1990
89   Interview with the Recycling Development Officer for the London Borough of Camden, Mike Newport, 15 June 1990
90   Interview with the Recycling Development Officer for the London Borough of Camden, Mike Newport, 15 June 1990
91   Interview with the Highways and Cleansing Officer for Bethnal Green Neighbourhood, London Borough of Tower Hamlets, Gery McCleary, 11 July 1990
92   Financial Times Survey (1991) *Waste Management*, 26 November
93   Miller, K (1990) 'Waste not want not' *The Guardian*, 10 October, quoted in Gandy, M (1993) op cit, p 177
94   London Research Centre (1988) *London in Need* LRC, London; The London and South East Regional Planning Conference (1988) *Waste Recycling: A Regional Perspective* SERPLAN, London
95   Financial Times Survey (1991) *Waste Management*, 26 November
96   UK Department of the Environment (1990) *This Common Inheritance* HMSO, London. Thomas, D (1990) 'Tenfold rise sought for renewable energy' *The Financial Times*, 1 October
97   East London Waste Authority (1989) *Waste Disposal Plan*. Greater London Council (1983) *Landfill sites: methane generation and recovery*
98   *WARMER Bulletin*, August 1991
99   Bidwell, R and Mason, S (1975) 'Fuel from London's Refuse: An examination of economic viability' paper presented to the *National Conference on Conversion of Refuse to Energy* Montreux 3–5 November. Gulley, B (1979) 'Refuse disposal: Waste treatment systems – options for the 1980s' *Municipal Engineering*, 14 August, pp 600–604
100 Gandy, M (1993) op cit
101 South East London Combined Heat and Power Limited (no date) *SELCHP: A new direction in waste disposal for London*
102 Interview with the Waste Reduction Officer for the London Waste Regulation Authority, Jeff Cooper, 28 September 1992
103 *The Financial Times*, 6 August 1990
104 *Planning*, 16 August 1991
105 Porteus, A (1990) 'Municipal Waste Incineration in the UK – What's Holding It Back' *Environmental Health*, pp 181–186
106 Brown, M (1991) 'More than a load of old rubbish' *The Times*, 14 January
107 *The Financial Times*, 4 October 1990
108 UK Department of the Environment (1990) *This Common Inheritance* HMSO, London

NEW YORK

1 Young (1991) *Discarding the Throwaway Society* Worldwatch Institute, Washington, DC
2 One closed section of the Fresh Kills landfill stands at 210ft compared with 204 ft for the Mycerinus (Menkaure) pyramid, while it is estimated that by the year 2010 one active section of the site will reach 450 ft, exceeding the height of the King Kephren pyramid (telephone interview with Nick Dmytryszyn, Environmental Engineer to Staten Island President, 18 October 1993; Information on the Egyptian Pyramids from the Egyptian Tourist Office, London)
3 Goldstein, E A and Izeman, M A (1990) *The New York Environment Book* Natural Resources Defense Council, Island Press, Washington, DC
4 Melosi, M (1981) *Garbage in the Cities: Refuse Reform and the Environment* Texas A&M University Press
5 Soper (1909) *Modern methods of street cleaning* Archibald Constable and Company Ltd, op cit, pp 161–162
6 Melosi, M (1981) op cit
7 Soper (1909) op cit
8 Soper (1909) op cit p 175
9 Waring, G E (1896) *A Report on the Final Disposition of the Wastes of New York by the Department of Street Cleaning* Martin B Brown, New York
10 Soper (1909) op cit
11 Melosi (1981) op cit p 179
12 Rogus, C A (1955) 'New York City makes team-mates of ... Sanitary Fills and Incinerators' *American City*, March, pp 114–115
13 Soper (1909) op cit
14 City of New York Department of Sanitation (1984) *The Waste Disposal Problem in New York City: A Proposal for Action* submitted in April 1984
15 Rogus, C A (1955) op cit
16 Melosi (1981) op cit
17 Goodrich, E P 1935 'The Opportunities for Refuse Salvage in New York City' *American City* February, p 58; City of New York Department of Sanitation (1992) *A Comprehensive Solid Waste Management Plan for New York City and Final Generic Environmental Impact Statement* submitted in August 1992
18 Tannenbaum, S (1992) *A brief history of waste disposal in New York City since 1930* unpublished paper held in the New York City municipal archives
19 Rogus, C A (1955) op cit
20 City of New York Department of Sanitation (1985) *Final Environmental Impact Statement for Proposed Resource Recovery Facility at the Brooklyn Navy Yard* submitted in June 1985
21 City of New York Department of Sanitation (1992) *A Comprehensive Solid Waste Management Plan for New York City and Final Generic Environmental Impact Statement* submitted in August 1992
22 Melosi, M (1981) op cit
23 'Town Says No to New York Garbage' *Governing*, May 1989 p 19; 'Garbage Is One Thing But Garbage From New York? Forget It' *New York Times*, 12 February 1989
24 City of New York Department of Sanitation (1984) *The Waste Disposal Problem in New York City: A Proposal for Action* submitted in April 1984

25   'Law is Urged to Require Drink-Container Deposits' *New York Times*,
     4 March 1975
26   'What this state needs is a good 5 cent returnable bottle' *The West Sider*,
     23 September 1976
27   New York State (1981) *'Inconspicuous Consumption: A look at New York's
     Bottle Bill and the experience of the six states which have enacted returnable contain-
     er legislation'* report by Assemblyman G Oliver Koppell of Bronx County,
     April 1981
28   Figures quoted in the *New York Times*, 5 October 1992
29   Interview with the Chief of Recycling Operations for the New York City
     Department of Sanitation, Anthony Lofaso, 2 September 1993
30   Report of the Moreland Act Commission on the Returnable Container Act
     submitted in March 1990 to Governor Mario Cuomo of New York State
31   'Can Picker: $35 a Shift, No benefits, No bosses' *New York Times*,
     6 September 1989
32   City of New York Department of Sanitation (1992) *Public Comments on A
     Comprehensive Solid Waste Management Plan for New York City and Draft Generic
     Environmental Impact Statement* submitted in June 1992
33   'A Middleman's Ventures in the Can Trade' *New York Times*,
     23 September 1992
34   Report of the Moreland Act Commission on the Returnable Container Act
     submitted in March 1990 to Governor Mario Cuomo of New York State
35   City of New York Department of Sanitation (1992) *Public Comments on A
     Comprehensive Solid Waste Management Plan for New York City and Draft Generic
     Environmental Impact Statement* submitted in June 1992
36   Interviews with Nancy Wolf and Steve Richardson of the Environmental
     Action Coalition, New York City, 26 August 1993
37   Interview with the Chief of Recycling Operations for the City of New York
     Department of Sanitation, Anthony Lofaso, 2 September 1993
38   'For "Recycling Cops", Dragnets Turn Up Bottles and Cans' *New York Times*,
     23 November 1990; 'Recycling Cops begin writing tickets on Island' *Staten
     Island Advance*, 1 September 1990; 'Mandatory recycling law offenders get
     summons' *New York Amsterdam News*, 1 September 1990
39   City of New York Department of Sanitation (1992) *A Comprehensive Solid
     Waste Management Plan for New York City and Final Generic Environmental
     Impact Statement* submitted in August 1992; Interview with the Outreach Co-
     ordinator for the Bureau of Waste Prevention, Reuse and Recycling in the
     City of New York Department of Sanitation, Patricia Grayson, 26 August 1993
40   Site visit to the East Harlem materials recovery facility with the Project
     Manager for the Contracts Unit of the City of New York Department of
     Sanitation, Penny Miller, 3 September 1993
41   Site visit to the East Harlem materials recovery facility with the Project
     Manager for the Contracts Unit of the City of New York Department of
     Sanitation, Penny Miller, 3 September 1993
42   City of New York Office of the Comptroller (1993) *The trash is always greener
     on the other side: A plan to keep recycling jobs on the New York City side of the
     Hudson* submitted in August 1993
43   See for example Castells, M (1989) *The Informational City*; Sassen, S (1991) *The
     Global City: New York, London, Tokyo* Princeton University Press, Princeton, NJ
44   City of New York Office of the Comptroller (1993) *The trash is always greener
     on the other side: A plan to keep recycling jobs on the New York City side of the
     Hudson* submitted in August 1993

45 'Recycling Gets Reprieve and Reproof' *New York Newsday*, 30 October 1991
46 'Budget Woes May Mean Dirty Streets Dirty City' *New York Observer*, 13 May 1991; 'Outlook Grim and Dirty For New York Ecology' *New York Times*, 16 July 1991
47 'Recycling Programme Ordered Expanded – Failure to Fund Project Deemed Immaterial' *New York Law Journal*, 5 February 1992
48 'The Recycling Blues' *New York Advertiser* 13 September 1992; Aquino, J T and White, K M (1993) 'Am I Blue? A Report on Blue Bag Recycling Programs' *Waste Age* July, pp 29–36
49 'Recycling is costing us millions' *Staten Island Advertiser*, 1 April 1990; 'Recycling plan cost will soar says report' *New York Daily News*, 12 October 1990; 'Cost of recycling trash more than city expected' *Staten Island Advance*, 25 February 1991
50 Starr, R (1991) 'Waste Disposal: a miracle of immaculate consumption?' *Public Interest*, p 26
51 Natural Resources Defense Council (1992) *Waste Watch 2nd Annual Report on New York City's Mandatory Recycling Programme* October 1992, NRDC, New York City; 'Judge Rebukes City For Recycling Failure' *New York Newsday*, 19 February 1992; 'Double Efforts on Recycling City Ordered' *New York Newsday*, 3 March 1992
52 'New York City Will Confront Mafia in Chinatown by bidding to Haul Garbage of Some Businesses' *Wall Street Journal*, 25 April 1988; Interviews with Nancy Wolf and Steve Richardson of the Environmental Action Coalition, New York City, 26 August 1993; Interview with the Chief of Recycling Operations for the City of New York Department of Sanitation, Anthony Lofaso, 2 September 1993
53 'Environmentalists fear a recycling breakdown ' *Staten Island Sunday Advance*, 27 January 1991
54 'Norman Steisel and the art of the done deal' *Village Voice*, 26 November 1991; 'Holzman: Let's Burn City's Incinerator Plan' *New York Newsday*, 16 January 1992
55 'We're on the cutting edge of compost' *New York Newsday*, 6 March 1993; Interviews with Nancy Wolf and Steve Richardson of the Environmental Action Coalition, New York City, 26 August 1993
56 Citywide Recycling Advisory Board (1991) *Recycle First: A Solid Waste Management Plan for New York City* submitted in December 1991
57 City of New York Office of the Comptroller (1992) *What Goes Around Comes Around: Good news about recycling markets* submitted in June 1992
58 City of New York Department of Sanitation (1991) *Preliminary Recycling Plan – Fiscal Year 1991*, p 2
59 'The Poor Mainly Recycle Poverty' *New York Times*, 30 December 1990; 'Recycling Facing Test in Poorer Areas' *New York Times*, 23 September 1992
60 'Ardent Recyclers meet Grim Realists – Superintendents from Bronx Apartment Buildings Say They Know the Hard Facts' *New York Times*, 14 November 1992. Interviews with the Outreach Co-ordinator for the Bureau of Waste Prevention, Reuse and Recycling in the City of New York Department of Sanitation, Patricia Grayson, and with the Assistant to the Director of Recycling Implementation, Charles 'Chuck' Arnold, on 26 August 1993
61 Interview with the Outreach Co-ordinator for the Bureau of Waste Prevention, Reuse and Recycling in the New York City Department of Sanitation, Patricia Grayson, 26 August 1993; 'Waste They Want Not in

Jackson Avenue: Area's up in arms over transfer station plan' *New York Newsday*, 6 May 1991; 'Brooklyn Queens Fighting Garbage Transfer Sites' *New York Newsday*, 29 June 1991

62   Interview with the Outreach Co-ordinator for the Bureau of Waste Prevention, Reuse and Recycling in the City of New York Department of Sanitation, Patricia Grayson, 26 August 1993

63   City of New York Department of Sanitation (1984) *The Waste Disposal Problem in New York City: A Proposal for Action* submitted in April 1984, p 1–6

64   Goldstein, E A and Izeman, M A (1990) *The New York Environment Book* Island Press, Washington, DC; 'Norman Steisel and the art of the done deal' *Village Voice*, 26 November 1991

65   Interview with the Assistant to the Director of Recycling Implementation City of New York Department of Sanitation, Charles 'Chuck' Arnold, 26 August 1993

66   Interviews with Nancy Wolf and Steve Richardson of the Environmental Action Coalition, New York City, 26 August 1993

67   de Kadt, M and Lilienthal, N (1989) *Solid Waste Management: The Garbage Challenge for New York City* INFORM, New York City; City of New York Office of the Comptroller (1992) *Burn, Baby, Burn: How to dispose of Garbage by Polluting Land Sea and Air at Enormous Cost* submitted in January 1992

68   City of New York Department of Sanitation (1984) *The Waste Disposal Problem in New York City: A Proposal for Action* submitted in April 1984

69   See, for example, Shefter, M (1985) *Political Crisis/Fiscal Crisis* Basic Books, New York; Bailey, R W (1984) *The Crisis Regime – the New York City Financial Crisis* State University of New York Press Albany; City of New York Department of Sanitation (1984) *The Waste Disposal Problem in New York City: A Proposal for Action* submitted in April 1984

70   City of New York Department of Sanitation (1992) *A Comprehensive Solid Waste Management Plan for New York City and Final Generic Environmental Impact Statement* submitted in August 1992

71   'Norman Steisel and the art of the done deal' *Village Voice*, 26 November 1991, p 37. See also Steisel, N and Casowitz, P (1984) 'The waste disposal crisis and promise of resource recovery' *New York Affairs*, 8, pp 105–116

72   'Dinkins Burning for Incinerator' *New York Newsday*, 26 August 1992; 'Full steam ahead for Garbage Plan' *New York Newsday*, 27 August 1992

73   New York State Department of Environmental Conservation Division of Solid and Hazardous Waste (1987) *Ash Residue Characterization Report* submitted in July 1987

74   City of New York Department of Sanitation (1992) *Public Comments on A Comprehensive Solid Waste Management Plan for New York City and Draft Generic Environmental Impact Statement* submitted in June 1992

75   City of New York Office of the Comptroller (1992) *Burn, Baby, Burn: How to dispose of Garbage by Polluting Land Sea and Air at Enormous Cost* submitted in January 1992. City of New York Office of the Comptroller (1992) *Smokescreen: How the Department of Sanitation's Solid Waste Plan and Environmental Impact Statement cover up the poisonous health effects of burning garbage* submitted in June 1992

76   City of New York Office of the Comptroller (1992) *Burn, Baby, Burn: How to dispose of Garbage by Polluting Land Sea and Air at Enormous Cost* submitted in January 1992

77   Environmental Defense Fund (1985) *To Burn or Not to Burn: The economic advantages of recycling over garbage incineration for New York City* EDF, New

York City; New York Public Interest Research Group (1992) *A Fiscal Analysis of the City of New York's Solid Waste Management programs and the Proposed Brooklyn Navy Yard Incinerator* NYPIRG, New York City

78 City of New York Office of the Comptroller (1992) *Fire and Ice: How Garbage Incineration Contributes to Global Warming,* submitted in March 1992

79 City of New York Office of the Comptroller (1992) *A Tale of Two Incinerators: How New York City Opposes Incineration in New Jersey While Supporting It At Home* May 1992

80 Natural Resources Defense Council (1992) *Statement of Eric Goldstein of the Natural Resources Defense Council before the New York City Council Environmental Protection Committee concerning, New York City's Solid Waste Management Plan* 25 August 1992, NRDC, New York City; City of New York Office of the Comptroller (1993) *The trash is always greener on the other side: A plan to keep recycling jobs on the New York City side of the Hudson* submitted in August 1993

81 City of New York Department of Sanitation (1992) *Public Comments on A Comprehensive Solid Waste Management Plan for New York City and Draft Generic Environmental Impact Statement* submitted in June 1992

82 US Environmental Protection Agency (1991) Risk Ranking Work Group Region II. *Overview Report: Comparative Risk Ranking of the Health Ecological and Welfare Effects of Twenty-Seven Environmental Problem Areas in Region II* EPA, Washington, DC

83 City of New York Department of Sanitation (1992) *A Comprehensive Solid Waste Management Plan for New York City and Final Generic Environmental Impact Statement* submitted in August 1992

84 City of New York Department of Sanitation (1992) *A Comprehensive Solid Waste Management Plan for New York City and Final Generic Environmental Impact Statement* submitted in August 1992, p 1–3

HAMBURG

1 Hartwich H-H (1990) *Freie und Hansestadt Hamburg: Die Zukunft des Stadtstaates* Landeszentrale für politische Bildung, Hamburg

2 Soper, GA (1909) *Modern methods of street cleaning* Archibald Constable and Company Ltd, London, p 121. For further details of the historical emergence of waste management in Hamburg see Mischer, O (1991) *Umweltgeschichte in Hamburg: Hausmüllentsorgung von Mitte der 20er bis Ende der 50er Jahre* magister dissertation at the University of Hamburg Frilling, H (1991) *Umweltgeschicte in Hamburg: Die Hausmüllentsorgung von 1886 bis 1960* magister dissertation at the University of Hamburg

3 Soper, G A (1909) op cit p 129

4 Clinker is the term used for the stony residue from burnt coal

5 Soper, G A (1909) op cit p 132

6 Meyer, F A (1901) *Die städtische Verbrennungsantalt für Abfallstoffe am Ballerdeich in Hamburg* Friedrich Wieweg & Sohn, Braunschwieg

7 Gandy, M (1993) *Recycling and Waste: An exploration of contemporary environmental policy* Avebury, Aldershot

8 Interview with Dr Eckhart Willing, Umweltbundesamt, Berlin, 16 September 1993

9 Ingenieurgemeinschaft Technischer Umweltschutz (1989) *Verfahren zur Reduktion des Hausmüllaufkommens* Ingenieurgemeinschaft Technischer Umweltschutz, Berlin

10   Interview with Wiebke Sager in the Amt für Entsorgungsplanung, Hamburg Baubehörde, 10 July 1990

11   Hedlund, A (1988) 'In Prosperity's Back Garden' paper presented to *Elmia Waste and Recycling Conference*, 13–17 June, Jönköping, Sweden

12   Interview with Dr Manfred Körner in the research department of the Hamburg branch of the Sozialdemokratische Partei Deutschlands, 4 July 1990

13   Interview with Wiebke Sager in the Amt für Entsorgungsplanung, Hamburg Baubehörde, 17 September 1990

14   Interview with Wiebke Sager in the Amt für Entsorgungsplanung, Hamburg Baubehörde, 17 September 1990

15   Interview with Martis Okekke in the Recycling Zentrum Harburg, 6 September 1990

16   Interview with Gerd Eich in the Landesbetrieb Hamburger Stadtreinigung, 5 September 1990

17   Bürgerschaft der Freien und Hansestadt Hamburg (1989) *Mitteilung des Senats an die Bürgerschaft* Drucksache 13/4091. Landesbetrieb Hamburger Stadtreinigung (1990) *Geschäftsbericht 1989* Umweltbehörde, Freie und Hansestadt Hamburg

18   Interview with Dr Ingo Wittern, Grüne Naßmüll Project Harburg, 6 September 1990. Interview with Martis Okekke, Recycling Zentrum Harburg, 6 September 1990

19   Interview with Gerd Eich in the Landesbetrieb Hamburger Stadtreinigung, 5 September 1990

20   Interview with Wiebke Sager, Amt für Entsorgungsplanung Hamburg Baubehörde, 10 July 1990

21   von Schönberg, A (1990) 'A Clear Issue: A survey of Glass Recycling in West Germany' *WARMER Bulletin*, 26, p 7; Interview with Dr Rohweder Landesbetrieb Hamburger Stadtreinigung, September 1993

22   Interview with Dr Rohweder, Landesbetrieb Hamburger Stadtreinigung, 15 September 1993

23   Interview with Gerd Eich in the Landesbetrieb Hamburger Stadtreinigung, 5 September 1990

24   Interview with Wiebke Sager in the Amt für Entsorgungsplanung, Hamburg Baubehörde, 10 July 1990

25   Freie und Hansestadt Hamburg (1989) *Abfallwirtschaftsplan* Baubehörde, Amt für Entsorgungsplanung, Hamburg

26   Interview with Gerd Eich in the Landesbetrieb Hamburger Stadtreinigung, 6 July 1990

27   Interview with Wiebke Sager in the Amt für Entsorgungsplanung, Hamburg Baubehörde, 17 September 1990

28   Interview with Martis Okekke, Recycling Zentrum Harburg, 6 September 1990

29   Interview with Dr Ingo Wittern, Grüne Naßmüll Project Harburg, 6 September 1990

30   Interview with Dr Ingo Wittern, Grüne Naßmüll Project Harburg, 6 September 1990

31   Interview with Michaela Moser, Institut für ökologisches Recycling, 20 September 1990

32   Interview with Wiebke Sager, Amt für Entsorgungsplanung, Hamburg Baubehörde, 17 September 1990

33 Ingenieurgemeinschaft Technischer Umweltschutz (1989) *Verfahren zur Reduktion des Hausmüllaufkommens* Ingenieurgemeinschaft Technischer Umweltschutz, Berlin

34 Interview with Wiebke Sager, Amt für Entsorgungsplanung, Hamburg Baubehörde, 17 September 1990

35 Interview with Dr Ingo Wittern, Grüne Naßmüll Project Harburg, 6 September 1990

36 Interview with Gerd Eich in the Landesbetrieb Hamburger Stadtreinigung, 5 September 1990

37 See, for example, Brown, L R; Flavin, C; and Postel, S (1990) 'Picturing a Sustainable Society' in Brown, L R (ed), *State of the World* Earthscan, London, pp 173–190. Pollock, C (1987) *Mining Urban Wastes: The Potential for Recycling* Worldwatch Paper 76, Worldwatch Institute, Washington, DC

38 Bürgerschaft der Freien und Hansestadt Hamburg (1989) *Mitteilung des Senats an die Bürgerschaft* Drucksache 13/4091

39 Gandy, M (1993) op cit

40 Interview with Dr Manfred Körner in the research department of the Hamburg branch of the Sozialdemokratische Partei Deutschlands, 4 July 1990

41 Interview with Gebhard Kraft, CDU Fraktion Hamburg, 13 September 1990

42 Interview with Annette Hallerberg, FDP Fraktion Hamburg, 7 September 1990

43 Christlich-Demokratische Union (1990) *Zum Abfallwirtschaftsplan (Hausmüll)* Pressekonferenz der CDU-Bürgerschaftsfraktion, 25 May. Christlich-Demokratische Union (1990) *Mißmanagement bei der Müllentsorgung: Millionengebühren für Hamburgs Bürger* Pressekonferenz der CDU-Bürgerschaftsfraktion, 12 September. Freie Demokratische Partei (1989) *Zukunft Hamburg: Liberale Ecktpunkte* FDP, Hamburg

44 Interviews with Jürgen Hannert and Antje Möller-Bierman, Die Grünen/GAL Hamburg, 10 July 1990. Interview with Knut Sander, Ökopol GmbH, 12 September 1990. Interview with Anna Wispler in the Institut für ökologisches Recycling, 16 September 1993

45 Interview with Anna Wispler, Institut für ökologisches Recycling, Berlin, 16 September 1993

46 Sozialdemokratische Partei Deutschlands (1989) *Berliner Koalitionsvereinbarung zwischen SPD und AL vom 13. März 1989* SPD Berlin

47 Mintzel, A and Oberreuter, H (eds) (1990) *Parteien in der Bundesrepublik Deutschland* Bundeszentrale für politische Bildung, Bonn

48 'Hamburg poll humiliates main parties' *The Guardian*, 20 September 1993

49 Interview with Annette Hallerberg, FDP Fraktion Hamburg, 7 September 1990. Interview with Gerd Eich in the Landesbetrieb Hamburger Stadtreinigung, 5 September 1990

50 Christlich-Demokratische Union (1988) *Hamburgs Beitrag zum Schutz des Klimas* Pressekonferenz der CDU-Bürgerschaftsfraktion, 4 November

51 Christlich-Demokratische Union (1988) *Mülldeponien in Hamburgs Landschaftsschutzgebieten* Presskonferenz der CDU Bürgerschaftsfraktion, 4 October

52 Interview with Gebhard Kraft, CDU Fraktion Hamburg, 13 September 1990

53 Interview with Michaela Moser, Institut für ökologisches Recycling, 20 September 1990. Interview with Dr Jürgen Ossenbrügge in the Geography Department at the University of Hamburg, 5 September 1990

54 'Es geht als doch: Aus Müll wird Gold gemacht' *Tageszeitung Hamburg*, 7 July

55   Interview with Dr Rohweder, Landesbetrieb Hamburger Stadtreinigung, 15 September 1993
56   Figures quoted in the *WARMER Bulletin*, May 1993
57   Inteview with Knut Sander, Ökopol GmbH, 12 September 1990
58   Ahrens, A; Gonsch, V; Ohde, J et al (1990) *Müllkonzept* Bund für Umwelt-und Naturschutz (BUND), Hamburg
59   Interviews with Jürgen Hannert and Antje Möller-Bierman, Die Grünen/GAL Hamburg, 10 July 1990. Interview with Knut Sander, Ökopol GmbH, 12 September 1990. Interview with Knut Sander, Ökopol GmbH, 12 September 1990
60   Ingenieurgemeinschaft Technischer Umweltschutz (1989) *Verfahren zur Reduktion des Hausmüll aufkommens* ITU, Berlin; Thiel, W (1988) *Kosten/leistungs aspekte und geeignete Einsatzgebiete der getrennten Naßmüllsammlung und Kompstierung am Beispiel Hamburgs* unpublished thesis, University of Hamburg
61   Ahrens, A; Gonsch, V; Ohde, J; et al (1990) *Müllkonzept* Bund für Umwelt-und Naturschutz (BUND), Hamburg, in translation, quoted in Gandy, M (1993) op cit, p 233
62   Interview with Knut Sander, Ökopol GmbH, 12 September 1990
63   Interview with Karl-Heinz Senkpiel, A H Julius Rohde GmbH, 13 September 1990
64   Gandy, M (1993) op cit
65   Bund für Umwelt und Naturschutz Deutschland (1992) *Verpackungsflut ohne Ende? Verpackungsverordnung und "Duales System"* BUND, Landesverband Niedersachsen
66   Interview with Dr Rohweder, Landesbetrieb Hamburger Stadtreinigung, 15 September 1993; 'Der Hausmüll und die Unfähigkeit Probleme zu losen' in *Frankfurter Rundschau*, 12 March 1993. Hibbeln K (1992) 'The development of waste management in Hamburg' paper presented to the conference *Resource Recovery from Waste*, 21–24 September, Imola, Sweden
67   Flood, M (1993) 'Over-packaging or over-consumption?' *WARMER Bulletin*, 37, pp 16–17
68   Interview with Dr Eckhart Willing, Umweltbundesamt, Berlin, 16 September 1993
69   'DSD takes over chemicals industry packaging waste' *Environment Business*, 30 July 1993
70   Genillard, A (1993) 'Bonn proposes new recycling measure' *Financial Times*, 1 April. Genillard, A (1993) 'Too much of a good thing' *Financial Times*, 23 June
71   'Denmark Halts Plans to Import German Waste' *Toxic Trade Update* number 6.2, 1993. 'Stop the EC waste war' *Financial Times*, 1 July 1993. Carvel, J (1993) 'Bonn accused of littering Europe' *The Guardian*, 5 May. Lean, G (1993) 'Waste tax plan offers lifeline to recyclers' *The Observer*, 16 May
72   'Indonesia Fights Plastic Waste Invasion' *Toxic Trade Update*, number 6.1, 1992; 'Plastic Wastes Accumulate in Indonesian Ports' *Toxic Trade Update*, number 6.2, 1993. 'The "Green Dot" Encourages Waste Trade' *Toxic Trade Update*, number 6.2, 1993. Greenpeace Toxic Trade Campaign (1993) *Plastic Waste to Indonesia: The Invasion of the Little Green Dots* Greenpeace, Hamburg
73   'Recycling Loophole Looms Large' *Toxic Trade Update*, number 6.1, 1993
74   'German packaging law condemned' *The Financial Times* 3 October 1991
75   Flood, M (1993) 'Over-packaging or over-consumption?' *WARMER Bulletin*, 37, pp 16–17

76   Akkre, E (1991) *Thermal recovery – part of an integrated waste management strategy* Alliance for Beverage Cartons and the Environment, London. Rassmussen, M (1991) *Waste to energy technology* Alliance for Beverage Cartons and the Environment, London
77   European Commission (1991) *Outline proposal for a Council Directive on Packaging* XI/270/91, European Commission, Brussels
78   'Cloak of subsidiarity is used to keep Britain polluting' *The Observer*, 6 December 1992
79   Bund für Umwelt und Naturschutz Deutschland (1992) *Irrwegstatt Ausweg*, BUND, Berlin; Bund für Umwelt und Naturschutz Deutschland (1992) *Der Müllpunkt* BUND, Berlin. Interview with Ute Dibbert, Bund für Umwelt und Naturschutz Deutschland eV, Berlin, 13 September 1993
80   Interview with Dr Eckhart Willing, Umweltbundesamt, Berlin, 16 September 1993
81   Entsorga GmbH (1993) *Das Duale System zum Erfolg führen* Dokumentation zum Entsorga-Congress, 1 July, Köln; Arnin Rockholz of the German Chamber of Commerce, address to the *20/20 Vision Conference* held at the Goethe Institute, London 8 July 1993

CONCLUSION

1   See, for example, Abert, J G (1985) *Municipal Waste Processing in Europe: A Status Report on Selected Materials and Energy Recovery Projects* The World Bank, Washington, DC
2   Gandy, M (1992) *The Environmental Debate: A Critical Overview* Geography Discussion Papers, University of Sussex
3   Gandy, M (1993) *Recycling and waste: an exploration of contemporary environmental policy* Avebury, Aldershot
4   McClaren, D (1992) 'London as Ecosystem' in Thornley, A (ed) *The Crisis of London* Routledge, London and New York, pp 56–69; Pollock, C (1987) *Mining Urban Wastes: The Potential for Recycling* Worldwatch Institute, Washington, DC; Young, J (1991) *Discarding the Throwaway Society* Worldwatch Institute, Washington, DC
5   See, for example, New York Public Interest Research Group (1992) *Setting the Record Straight: A Fiscal Analysis of the City of New York's Solid Waste Management Programs and the Proposed Brooklyn Navy Yard Incinerator* NYPIRG, New York
6   British Association of Nature Conservationists (1990) *The Conservationists' Response to the Pearce Report* BANC, London, p 23
7   See Institut für ökologisches Recycling (1991) *Perspektive Abfallvermeidung* IföR Berlin; Die Grünen (1989) *Verzicht auf PVC und Chlorchemie* Die Grünen, Bonn
8   Environmentalist calls for zero growth in the 1970s met with fierce opposition on the grounds that the social consequences would be intolerable. See, for example, Beckerman, W (1974) *In Defence of Economic Growth* Jonathan Cape, London
9   See, for example, Akkre, E (1991) *Thermal recovery – part of an integrated waste management strategy* Alliance for Beverage Cartons and the Environment, London
10 'Government to amend recycling targets' *Environment Business*, 16 June 1993

Further reading

This selected bibliography provides examples of recycling and waste management literature from a variety of different perspectives. Note that there is no consensus over either the cause of low levels of recycling or over the most appropriate approach to urban waste management:

Abert, J G (1985), *Municipal Waste Processing in Europe: A Status Report on Selected Materials and Energy Recovery Projects*, The World Bank, Washington DC.

Alter, H (1991), The Future Course of Solid Waste Management in the US, *Waste Management & Research*, 9, pp 3–20.

Arbeitsgemeinschaft ökologische Forschungsinstitute (1990), *Abfallvermeidung*, Raben Verlag, Munich.

Barton, A F M (1979), *Resource Recovery and Recycling*, John Wiley, New York.

Blumberg, L and Gottlieb, R (1989), *War on Waste: Can America Win Its Battle With Garbage*, Island Press, Washington, DC.

Boustead, I (1989), *The Environmental Impact of Liquid Food Containers in the UK*, Open University, Milton Keynes.

Brisson, I (1992), *Packaging Waste and the Environment: Economics and Policy*, CSERGE Report 91–01, University College London and University of East Anglia.

Bundesamt für Umwelt Wald und Landschaft (1991), *Oekobilanz von Packstoffen*, BUWAL, Bern.

Butlin, J A (1977), 'Economics and Recycling', *Journal of Resources Policy*, pp 87–95.

Cargo, D B (1978), *Solid Wastes: Factors influencing generation rates*, Geography Research Paper No 174, University of Chicago.

Chandler, W U (1983), *Materials Recycling: The Virtue of Necessity*, Worldwatch Paper 56, Worldwatch Institute, Washington, DC.

Clark, M, Smith, D and Blowers, A T (eds) (1992), *Waste location: Spatial aspects of waste management, hazards and disposal*, Routledge, London and New York.

Cohen, N, Herz, M, and Ruston, J (1988), *Coming Full Circle: Successful Recycling Today*, Environmental Defense Fund, New York.

Cointreau, S J, Gunnerson, C G, Huls, J M, and Seldman, N N (1984), *Recycling from Municipal Refuse: A State-of-the-Art Review and Annotated Bibliography*, the World Bank, Washington, DC.

Cointreau, S J (1987), *Solid Waste Recycling: Case Studies in Developing Countries*, The World Bank, Washington, DC.

Cubbin, J, Domberger, S, and Meadowcroft, S (1988), 'Competitive Tendering and Refuse Collection: Identifying the Sources of Efficiency Gains', *Fiscal Studies*, 8, pp 49–58.

Curlee, T R (1986), *The economic feasibility of recycling: A case study of plastic wastes*, Praeger Publishing, New York.

Darnay, A and Franklin, W E (1972), *Salvage Markets for Materials in Solid Wastes*, US Environmental Protection Agency, Washington, DC.

David Perchard Associates (1993), *EC Directive on Packaging and Packaging Waste: Compliance Cost Assessment.*, David Perchard Associates, London.

De Young, R (1986), 'Some Psychological Aspects of Recycling: The Structure of Conservation Satisfactions', *Environment and Behaviour*, 18, pp 435–449.

Denison, R A and Ruston, J (1990), *Recycling and Incineration: Evaluating the Choices*, Island Press, Washington, DC.

Domberger, S, Meadowcroft, S and Thompson, D (1988), 'Competitive Tendering and Efficiency: The Case of Refuse Collection', *Fiscal Studies*, 7, pp 69–87.

Dourado, P (1990), 'The case against recycling the US's waste', *New Scientist*, 8 September, p 48.

The Economist (1991), 'Recycling in Germany: a wall of waste', 30 November, p 97.

The Environmental Defense Fund (1985), *To Burn or Not to Burn: The Economic Advantages of Recycling Over Garbage Incineration for New York City*, EDF, New York.

The Environmental Defense Fund (1987), *Coming Full Circle: Successful Recycling Today*, EDF, New York.

European Commission (1984), *Employment Potential of Waste Recovery and Recycling Activities and Socio-Economic Relevance of Waste Management Sector in the Community*, final report by Environment Resources Ltd for the Directorate General for the Environment.

European Commission (1985), *The Employment Implications of Glass Re-use and Recycling*, Report by Ecotec Research and Consulting Ltd for the Directorate General for Science, Research and Development.

European Commission (1989), *A Community Strategy for Waste Management*, SEC(89)934, European Commission, Brussels.

European Commission (1991), *Outline proposal for a Council Directive on Packaging*, XI/270/91, European Commission, Brussels.

European Commission (1991), *Proposal for a Council Directive on the landfill of waste*, COM (91) 102, European Commission, Brussels.

Everett, J W (1989), Residential recycling programs: environmental, economic and disposal factors, *Waste Management & Research*, 7, pp 143–152.

Financial Times Survey (1991), *Waste Management*, 26 November.

Financial Times Survey (1992), *Environmental Management*, 10 November.

Friends of the Earth (1992), *Don't Throw it All Away! Friends of the Earth's Guide to Waste Reduction and Recycling*, Friends of the Earth, London.

Gandy, M (1993), 'A critical analysis of environmental policy in developed economies: the case of recycling', in *Policies, Institutions and the Environment*, pp 1–29, South North Centre for Environmental Policy, School of Oriental and African Studies, University of London.

Gandy, M (1993), *Recycling and Waste: an exploration of contemporary environmental policy*, Avebury, Aldershot.

Gardner, D, Peel, Q, and Hunt, J (1992), 'Green Germany drags Brussels into environmental arena', *The Financial Times*, 24 January.

Gatrell, A C and Lovett, A A (1992), 'Burning questions: incineration of wastes and implications for human health', in Clark, M, Smith, D and Blowers, A (eds),

Spatial aspects of waste management, hazards and disposal, Macmillan, London, pp 143–158.

Genillard, A (1993), 'Bonn proposes new recycling measure', *Financial Times*, 1 April.

Genillard, A (1993), 'Too much of a good thing', *Financial Times*, 23 June.

Goddard, H C (1975), *Managing Solid Wastes: Economics, Technology and Institutions*, Praeger, New York.

Gore, A (1992), *Earth in the Balance: Forging a new common purpose*, Earthscan, London. (See the chapter on waste.)

Hay, A, Wright, G, and Forshaw, J (1990), *Fashionable Waste: the make-up of a recycler*, Save Waste and Prosper Ltd, Leeds.

Hayes, D (1978), *Repairs, Reuse, Recycling – First Steps Toward a Sustainable Society*, Worldwatch Paper 23, Worldwatch Institute, Washington, DC.

Henstock, M E (1988), *Design for Recyclability*, Institute of Metals, UK.

Hockley, G C, Walters, J, and Goodall, P (1989), *Generating Profit from Waste*, Special Report No.1182, The Economist Intelligence Unit, London.

Hughes, D (1974), 'Towards a recycling society', *New Scientist*, 10 January, pp 58–60.

Institut für ökologisches Recycling (1988), *Abfall Vermeiden*, Fischer Taschenbuch Verlag, Frankfurt am Main.

Institut für ökologisches Recycling (1989), *Ökologische Abfallwirtschaft: Umweltvorsage durch Abfallvermeidung*, IföR, Berlin.

Institut für ökologisches Recycling (1991), *Perspektive Abfallvermeidung*, IföR, Berlin.

Kharbanda, O P and Stallworthy, E A (1990), *Waste Management: Towards a Sustainable Society*, Greenwood Press, Westport, Connecticut.

Koch, T C, Seeberger, J, and Petrik, H (1986), *Ökologische Müllverwertung: Handbuch für optimale Abfallkonzepte*, C F Muller, Karlsruhe.

Kut, D and Hare, G (1981), *Waste Recycling for Energy Conservation*, The Architectural Press, London.

Lean, G and Ghazi, P (1992), 'Cloak of subsidiarity is used to keep Britain polluting', *The Observer*, 6 December.

Lean, G (1993), 'Waste tax plan offers lifeline to recyclers', *The Observer*, 16 May.

Melosi, M V (1981), *Garbage in the Cities: Refuse, Reform and the Environment 1880–1980*, Texas A&M University Press.

Newsday (1989), *Rush to Burn: Solving America's Garbage Crisis?*, Island Press, Washington, DC.

Organisation for Economic Cooperation and Development (1983), *Separate Collection and Recycling*, OECD, Paris.

Packard, V (1960), *The Wastemakers*, David McKay Company Inc, London.

Page, T (1977), *Conservation and Economic Efficiency*, John Hopkins University Press, Baltimore.

Pieters, R G M and Verhallen, T M M (1986), 'Participation in Source Separation Projects: Design characteristics and percieved costs and benefits', *Resources and Conservation*, 12, pp 95–111.

Platt, B (1990), *Beyond 40%: Record-Setting Recycling and Composting Programs*, Insitute for Self Reliance/Island Press, Washington , DC.

Pollock, C (1987), *Mining Urban Wastes: The Potential for Recycling*, Worldwatch Paper 76, Worldwatch Institute, Washington, DC.

Robinson, W D, *The Solid Waste Handbook: A Practical Guide*, John Wiley and Sons, New York.

Savas, E S (1979), 'Public versus Private Refuse Collection: A Critical Review of the Evidence', *Urban Analysis*, 6, pp 1–13.

Stölting, P and Rubik, F (1992), *Übersicht über ökologsiche Produktbilanzen*, Institut für ökologsiche Wirtschaftsforshung (IW), Heidelberg.

Thome-Kozmiensky, K J (ed) (1984), *Recycling International*, EF-Verlag für Umwelt und Technik, Berlin.

Thome-Kozmiensky, K J (ed) (1989), *Müllverbrennung und Umwelt 3*, EF-Verlag für Energie und Umwelttechnik, Berlin.

Turner, R K (1992), *An Economic Incentive Approach to Regulating the Throwaway Society*, CSERGE Working Paper PA 92–05, University of East Anglia and University College London.

UK Department of Trade and Industry (1992), *Economic Instruments and Recovery of Resources from Waste*, HMSO, London.

UK House of Commons Trade and Industry Committee (1985), *The Wealth of Waste*, Second Special Report from the Trade and Industry Committee Session 1984–85, HMSO, London.

Royal Commission on Environmental Pollution (1993), *Incineration of Waste*, HMSO, London.

Underwood, J D, Hershkowitz, A, and de Kadt, M (1988), *Garbage: Practices, Problems and Remedies*, Inform Inc, US.

Valette, J (1989), *The International Trade in Wastes: A Greenpeace Inventory*, Greenpeace, Washington, DC. See also the quarterly *Toxic Trade Update* published by Greenpeace Toxic Trade Campaign, 1436 U Street, NW, Washington, DC 20009.

Vidal, J (1992), 'The new waste colonialists', *The Guardian*, 14 February.

Vining, J and Ebreo, A (1990), 'What Makes A Recycler? A Comparison of Recyclers and Nonrecyclers', *Environment and Behaviour*, 22, pp 55–73.

Vogler, J (1981), *Work from Waste: Recycling wastes to create employment*, Oxfam and Intermediate Technology Publications Ltd, London.

The Women's Environmental Network (1989), *Dioxin: A Briefing*, WEN, London.

The Women's Environmental Network (1990), *UK Paper Mills: Environmental Impact*, WEN, London.

The Women's Environmental Network (1990), *A Tissue of Lies? Disposable Paper and the Environment*, WEN, London.

Wolf, N and Feldman, E (1991), *Plastics: America's Packaging Dilemma*, Island Press, Washington, DC.

Young, J E (1991), *Discarding the Throwaway Society*, Worldwatch Institute, Washington, DC.

Index